Maximizing Profitability with Safety Culture Development

Maximizing Profitability with Safety Culture Development

Clifford M. Florczak

An imprint of Elsevier Science
Amsterdam Boston London New York Oxford Paris San Diego
San Francisco Singapore Sydney Tokyo

Butterworth-Heinemann is an imprint of Elsevier Science.

Copyright © 2002, Elsevier Science (USA). All rights reserved.

No part of this publication may be reproduced, stored in a retrieval system, or transmitted in any form or by any means, electronic, mechanical, photocopying, recording, or otherwise, without the prior written permission of the publisher.

♾ Recognizing the importance of preserving what has been written, Elsevier Science prints its books on acid-free paper whenever possible.

Library of Congress Cataloging-in-Publication Data

Florczak, Clifford M., 1951–
 Maximizing profitability with safety culture development / Clifford M. Florczak.
   p. cm.
 Includes index.
 ISBN 0-7506-7610-8 (alk. paper)
  1. Industrial safety—Economic aspects—United States. 2. Industrial hygiene—Economic aspects—United States. 3. Profit—United States. I. Title.

HD 7654 .F57 2002
658.3′82—dc21

2002028178

British Library Cataloguing-in-Publication Data
A catalogue record for this book is available from the British Library.

The publisher offers special discounts on bulk orders of this book.
For information, please contact:

Manager of Special Sales
Elsevier Science
200 Wheeler Road
Burlington, MA 01803
781-221-2212
781-221-1615

For information on all Butterworth-Heinemann publications available, contact our World Wide Web home page at: http://www.bh.com

10 9 8 7 6 5 4 3 2 1

Printed in the United States of America

Contents

1. Introduction **1**

Safety Culture, 1
The Need for Change, 6

2. Management Philosophies and Theory **10**

Accidents Are Like Icebergs, 12
(Not So) Modern Management Theories, 14
The Hawthorne Effect, 15
X, Y, Z Workers, 18
  Example: Unsafe Widget Stacking, 21
Why Workers Perform Unsafe Acts, 22
  Example: The Slippery Slope of Safety, 23
The Need for Culture: Selling the Program, 25
Summary, 26

3. Assessment Tools **28**

Accident Trends, 28
  OSHA Logs, 29
    Example: OSHA Log Compliance, 29
    Example: The Monday Morning Blues, 32
  First-Aid/Incident Logs, 35
  Individual Accident Reports, 36
  Accident Investigations, 37
  Medical Records, 38
Other Tools to Determine Safety Culture Development, 39
  Past OSHA Citations, 39
  Safety Meetings, 41
  Safety Inspections and Safety Audits, 41
  Job Safety Analyses, 41
  General Assessments, 41
Summary, 42

4. Program Implementation: How to Start **44**

Screening New Hires, 46
The Role of the Receptionist, 46

The Need for a Thorough Interview, 47
Multilanguage Applications, 49
Reference Checks, 50
Background Checks, 52
Working with Seasoned Veterans, 52
Summary, 53

5. Medical Surveillance/Drug Testing 55

Medical Surveillance, 56
   Lessons Learned: Preemployment Physicals, 57
   Lessons Learned: Medical Information Sharing, 60
   Dealing with Disgruntled Employees, 62
Drug Testing, 63
   Lessons Learned: Surprising Drug Screen Results, 64
Drug Screening and Timing, 65
Summary, 70

6. New Hire Orientation 71

Office Orientation, 72
Using Videos for Orientation, 73
   Example: Orientation for an On-site Visit, 79
The Use of Interactive Computer Orientation, 80
Summary, 81

7. Training 83

How Training Affects Safety Culture, 84
   Example: All Workers Need Training, 84
The Job Safety Analysis, 85
   Example: Training and Safety in the Front Office, 88
The Basic Steps of Training, 89
   Determining the Training Objectives, 89
      Example: Too Much of a Good Thing, 90
      Example: Going Around the System, 92
   Determining the Who, What, Where, and How of the Training, 97
   Determining if the Training Met the Objectives, 102
   Reviewing, Assessing, and Adjusting the Training, 103
Drivers and Providers of Training, 105
Summary, 108

8. Safety Performance as Part of Performance Evaluations 111

Performance Evaluations, 111
   Example: Safety and Performance, 113

Example: Accident-Prone Pete, 114
Accident Prone, 120
The Link Between Safety Performance and Performance
   Evaluations, 121
Should Safety Be Related to General Performance?, 121
Should Safety Be Linked to Performance Evaluations?, 122
   Myth: Union Workers Should Not Receive Performance
      Evaluations, 122
Performance Evaluation Content, 123
Summary, 125

9. Safety Incentive Bonus Programs 126

A Chain Is as Strong as Its Weakest Link, 128
   Management's Incentive, 131
   Hourly Incentive, 131
Cash Awards, 133
Group Safety Milestones, 133
   Example: The Job No One Wants, 133
   Lessons Learned: A Team Dissolved, 135
   Example: The Misplaced Celebration, 135
Individual Incentive Awards, 136
Summary, 137

10. Planning for Field Compliance 138

The Safety Program, 141
   Example: Exposure Limits, 142
Site-Specific Safety Plan, 143
   Example: An All-in-One Safety Plan, 148
   Example: When OSHA Comes Calling, 148
Summary, 149

11. Safety Culture Barometers 151

OSHA Logs and Other Accident-Related Information, 153
Trend Determination, 154
   Detailed Records Must Be Kept, 156
   Example: Statistics Aren't Always What They Seem, 157
   Example: Back Injuries, 157
Safety and Health Program Analysis Methods, 158
Perceived Obstacles to Safety, 162
Supervisor- and Employee-Identified Obstacles, 162
   Fear of Losing My Job, 162
   No Money for Needed Changes, 163
      Example: The Story of Chitty-Chitty Bang-Bang, 164
   No Support from Upper Management, 165
   Lack of Trust: Poor Ethics within Organization, 166

viii   Maximizing Profitability with Safety Culture Development

      Lack of Open Communication and Listening, 167
      Competing Priorities: Production Is Number One, 167
         Example: The Friendly Skies, 168
      Potential CEO and Top Management-Identified Obstacles, 169
      Summary, 170

12. Measuring Field Compliance 171

      Measuring Field Compliance Using Safety Inspections, 172
      Worker-Level Inspections, 172
         Housekeeping, 173
         Pretrip Inspections, 173
         A Closer Look at Professional Judgment, 178
           Example: The Little Cut That Became Something Bigger, 183
         "Machoism", 184
      Inspections at the Field Supervisor Level, 185
      Manager-Level Inspections, 187
      Measuring Field Compliance Using Safety Audits, 188
         Example: A Typical Safety Audit, 189
      What Triggers a Safety Audit, 190
      Getting an Audit Program Started, 190
         Example: Auditing Apples to Oranges, 194
      Ensure That the Audit is "Fair," 196
      Summary, 197

13. Increasing Program Effectiveness 199

      Safety Audit Review Boards, 200
      Facilitating Change through Programs, 202
      Choosing a Safety Audit Program, 204
         Example: Allegations Abroad, 205
      Making Reasonable Recommendations, 206
      Specific Case Analysis, 210
      Summary, 213

14. Training and Grooming Front-Line Supervisors 215

      How FLSs Are Chosen, 218
         Skill Level, Education, Training, and Experience, 218
         Ability to Accept Change, 219
         Leadership, 219
           Example: The Case of the Inflated Numbers, 221
         Mutual Respect, 222
         Ability to Communicate, 222
           Example: Creative Problem Solving, 224
         Human Resources, 226
      The "Gotcha" Attitude, 229

Example: Feuding Factory Workers, 229
The Supervisor's Role in Safety Inspections, 231
The Supervisor's Role in Accident Investigations, 233
The Late Reporting of Incidents, 235
FLS Postaccident Investigation Involvement, 237
  Example: Getting Official Medical Approval, 238
  Example: Abiding by the Doctor's Orders, 239
Working Within Restrictions, 241
  Example: Rushing the Healing Process, 243
The Supervisor's Involvement With JSAs, 245
Personal Protective Equipment, 245
Summary, 247

15. The Effects of a Tight Economy on Safety Programs and Culture 250

Safety Incentive Plans, 250
Elimination of Lucrative Safety Programs, 252
  Example: Underreporting Can Be Serious Business, 253
Safety Audits and Safety Inspections, 254
Safety Service Awards, 256
Safety Celebrations, 257
Summary, 258

Appendix A: Job Safety Analysis (JSA) 261

Appendix B: Tools for a Safety and Health Program Assessment 273

Appendix C: Draft Safety and Health Program Rule 291

Appendix D: Example Inspection Form 299

Appendix E: General Audit Form 311

Index 325

1

Introduction

Are you concerned over the rising costs of accidents and the effect that these types of losses can have on your company? If you are not, you should be. The cost of accidents and unplanned losses can drive a seemingly successful business out of business. But what can be done to eliminate or control accidents and losses? After all, accidents just happen—or do they?

The answer to that question is simple: Accidents don't just happen, they happen for a reason. Or, more accurately, they happen for a whole lot of reasons. Most people think very little can be done to prevent work-related accidents, but—to put it bluntly—they are wrong. A lot can be done to prevent accidents. The road to accident prevention may not be cheap, easy, or fast, but it is the right road to be on. This road is the development of safety culture.

SAFETY CULTURE

What is safety culture? When I refer to *safety culture* throughout this book, I am referring to how your company views and treats safety-related issues. The safety culture at the company where you work is the company's way of doing (safety-related) activities. It is an important part of how you and your company do business. When I refer to *safety*, I am referring to the general field of safety, health, and on occasion, the environment.

Safety culture is part of your company's corporate or general culture, and every company has a unique culture. Even competitors that are putting out similar products have different company cultures. Sometimes these cultures are vastly different. If we look at typical U.S. values, some find it hard to believe that women in Afghanistan are treated as second-class citizens (at least from a Western point of view). How is it that women cannot walk down the street by themselves and have to cover their faces? People in the West have a hard time understanding this type of so-called repressive culture.

When I try to imagine how the Afghan people look at the U.S. culture, I surmise that the Afghans are likely in shock that any country (such as the United States) could have a culture that does *not* include the (repressive) treatment of women as such. Most people in the United States are shocked when they see or hear about the (repressive) treatment of women, lack of allowance of music, and other policies. We believe that the Afghan way of doing things, or their culture, is just not "right."

I am confident that, conversely, many of the people of Afghanistan are shocked when they think about the freedoms that we have enjoyed in this country. I am sure that many Afghans believe that the way in which the United States does business—our "culture"—is just not "right." And if you are used to doing things a certain way and have certain beliefs (religious or otherwise), then you likely believe that your beliefs, and your way of doing business, is the "right" way.

Starting in the next chapter, a short discussion about theories of management is presented. If you are a management person at any level, I would like you to start asking yourself right now what you really think of your fellow workers, especially the hourly workers. Do you believe that people, workers, team members, and the like are basically "good," "bad," or none of the above? Is this goodness or badness learned or inherited? Do you think it can be changed? Do you think it *should* be changed? What do you think of your workforce in general?

Think of your company's culture. Think about how team members are treated. Are certain workers treated as second-class citizens? Does your company encourage openness? Does your company try to share pertinent company information with the workforce? Does your company share a concern for the safety, health, and well-being of team members?

What does your company do about reporting accidents? Is there an investigation associated with every accident, close call, first-aid case, and minor incident? What does your company do when a serious accident occurs? Is what happens when a serious incident occurs much different than when a close call or first-aid case occurs?

What is the benefit program like at your company? Are there provisions in your benefit program that include on-the-job and off-the-job coverage of activities? What did your company do when the team reached a milestone? How often does your company have a safety celebration? Do you or the rest of the team members believe that the company has integrity? Are all of the managers at your company in agreement and "walking the talk"?

The last few paragraphs were filled with a lot of questions. During the remainder of the book, we discuss these questions and the appropriate answers. If the answers you gave to any of these questions disappointed you, you might find a substantial amount of information offered within this book to be of interest.

You may believe the following: "My company is in a dangerous business. The hazards my co-workers deal with are real and cannot be avoided. There is no way to substantially improve." I have heard this response several times and believe that, in most cases, room for vast improvement probably does exist. Of course, this improvement does not come overnight and will involve time and effort on many people's part. After all, I am talking about changing culture, which is not a small task. I believe that changing culture is usually a monumental—but not insurmountable—task.

Here is another comment that seems to pop up often: "Why should we change? I think we do all right!" Unless your organization is not experiencing any accidents, I cannot agree with this statement; however, if your organization is perfect, there is no need to improve. In this case, the rest of this book will likely not interest you, so you can stop reading now!

Another statement that seems to be a subject of much discussion is: "I can't worry about accidents because I can't do anything about them. Besides, that's why we have insurance, isn't it?" Although having insurance is certainly important, this viewpoint seems to be missing the point completely. If we take an analogy, let's consider the person who believes that he has no need to wear a seatbelt while driving a car. After all, the vehicle that he drives has antilock brakes and side and front airbag systems. Let's look at some other situations: Why should one wear a hard hat when the person wearing it usually has a hard head? Why would anyone ever want to wear a seatbelt during takeoff and landing of a commercial flight? After all, if there is a crash, the seatbelt won't help anyway!

I have trouble agreeing with these statements. First, I have insurance, but I hope and pray that I never have to use it. I have met several victims after their residences have burned. I can't believe that any of them can say the insurance came close to replacing items that held high sentimental and emotional value—no amount of money can. Regarding wearing hard hats or any personal protective equipment, if the ensemble is chosen wisely, wearing it should be the prudent thing to do. I am not interested in allowing my hard head to contact a hazard. I would much rather have the hard hat perform in the manner for which it was designed.

Although I was not an eyewitness, I will never forget what I learned as part of the investigating team of a serious work-related accident. During the performance of an assigned work task, a team member wearing a hard hat was injured by falling materials, which struck him on the head. The hard hat was cracked, and the inner suspension was severed in two places. Basically, the hard hat was damaged, but it adequately performed the job it was assigned to do. The worker suffered some extensive injury, including head and neck trauma, fractured vertebrae, and so on, but I was amazed at the time of the accident that he did not die. I am confident he would have died if he had not worn that hard hat.

I agree that wearing safety equipment is usually not glamorous and typically not much fun. Let's take seatbelts as an example. Some people just will not wear their seatbelts. I had a co-worker who believed he had been saved from a car crash from which he was thrown clear. He was on his way home from work when his car left the pavement around a curve. During the accident, his car rolled over and was totaled. The driver's compartment was completely demolished. He came out of this accident with only minor injuries. He insisted that if he had been wearing his seatbelt, the accident would have killed him. Some people still believe this. There are also people who win the lottery, but I am not one of them.

By the way, this same person who swore that not wearing his seatbelt saved his life was killed less than a year later going home from the job on that same stretch of road. I know of very few people who can continue to believe that wearing a seatbelt is not a good idea when you drive. We now have many years of accident data to judge from. Those people boasting about being saved from potentially fatal crashes are those who have worn a seatbelt and not those who have not. Even if you have won the lottery, please wear your seatbelt while driving. Enough said.

How about another common statement I hear: "I would like to improve the safety culture, but I just don't think the organization is ready for another change." Business is always changing, and change is part of business. Not seizing the opportunity could be a big mistake. Maybe the change in safety culture would fit right in when combined with other changes. If we are rolling out another program, why not add to the agenda? The time for change is typically right now. If the time for change is not now, then when? Maybe the when is not here and now but will be the first of the year!

Planning the when for the *best* time is important, but putting off implementation indefinitely for lack of a good reason is not advisable. Figure out when, and prepare for implementation. Sometimes, when important things are delayed, one has a tendency to continue in that procrastination mode. Certainly, one can start planning for the changes now, and be ready when the right time comes.

Let's ask some more questions about our organization and the way the organization does business from a safety point of view. In other words, let's look at our own safety culture. When personal protective equipment (PPE) is deemed to be a requirement for certain work tasks, how does this PPE affect the team member? Does the company:

- expect the team member to buy it him or herself?
- begrudgingly agree to pay for it?
- use a subcontractor to avoid the situation entirely?
- comply with the union agreement?
- use another method?

There are many other ways to answer this question, and the answer serves as one of many small indicators of the status of your safety culture. The attitude and perception of how your company treats PPE is an important part of worker safety in many work environments.

Besides PPE, how about other aspects of safety culture? Management needs to consider how the organization keeps current with safety issues. Does the company have a staff or a lead person who is primarily responsible for guiding the company in safety-related areas? If the answer is yes, determine how this person gets trained and stays current. Does your organization allow for and encourage professional development? If so, does your company have a written policy? And is the policy adhered to? When was the last time this policy was reassessed? Currently, does this policy need to be reassessed? How often is someone allowed to go to professional development activities?

Each of these questions seems to lead to another question. The answers to the aforementioned questions and the basic theme of a corporate philosophy that will develop an effective safety culture lie in an organization that:

- cares about its people.
- cares about safety.
- has someone that team members know and respect who is the safety contact.
- trains and develops the safety people to keep knowledge and skill levels current.
- does whatever is necessary to provide a safe environment.

This list can be expanded considerably.

How about accidents? Is the organization concerned about industrial and other accidents? How about the Occupational Safety and Health

Administration (OSHA)? Is the organization concerned about a visit from OSHA? What transpired the last time OSHA inspected this facility? What happened the last time we had an OSHA inspection? What will happen to a worker if he or she complains about safety conditions? What will happen to a worker who calls OSHA? How about the recordability of accidents? Who decides this? Is there a modified-duty program for workers who are hurt on or off the job? Does the company bring people back from an accident on a stretcher if need be to make sure no time is lost from work?

How are accident investigations handled? Who handles them? Are the results of these investigations revealed to the workers? Does anyone learn from prior mistakes? What changes have taken place to make the facility safer? How does a team member know how to perform a job safely? Is there a safety manual? Does everyone who needs access to the safety manual have it? Is there a site-specific safety plan for current work activity? Who has access to this plan? How about job safety analyses (JSAs)? Have all of the work tasks been analyzed for safety concerns and appropriate safety systems been implemented? What does a worker do after a close call? Is there any investigation when no damage is done? Is any report made out?

All of these questions are answered throughout this book.

THE NEED FOR CHANGE

"How do I know the organization is ready for a change? Aren't we good enough?" If you are perfect, then you are good enough. If you are not perfect, then some adjustments should be considered. If your organization is ready for a change, there are probably some indicators of this. The forces that support change can come internally or externally.

Although the internal forces can vary, the healthiest situation is for top management to realize the need and implement improvements. Other scenarios from forces within the company do occur, but without buy-in from top management, implementing changes will be difficult, and ultimate success will likely not be forthcoming.

The external forces for change can vary substantially. An example of an external force that prompts a change in safety culture is regulatory bodies. Possibly OSHA or the state or local health department mandates that certain materials be treated in an upgraded manner. Failure to comply with these standards could result in a hefty fine. Certainly, no one would want to jeopardize their company by allowing the company to work in a way that may result in legal action or financial encumbrances.

Another likely external force is from outside competition. If your company performs work in a certain manner and a competitor provides this same service in a better, safer, cheaper, faster manner, then it is likely that over the long run you will lose business to your competitor. If this is the case, your competitors, in a way, are forcing you to provide improved services. Does your company strive to be an industry leader or to lag behind?

Developing or enhancing the current safety culture is a necessary task if your company wants to stay in business. Changing the safety culture will undoubtedly be a difficult and time-consuming task for a variety of reasons. Because culture is people-dependent, in order to build or develop culture, we must change people's attitudes and behaviors.

Workers at your company, like most people, will tend to resist change. Many workers are just plain satisfied and comfortable in what they do and the way they perform. People can get into a "groove" where they are doing similar things on a routine basis, with ease or very little effort or struggle. Some of you might consider this mode a "rut" rather than a "groove." But whether considered a rut or a groove, it is a way of life that is difficult to change.

Consider a work situation when dealing with the in-house phone or computer system: Think of the last time your company changed the way you needed to log on to the in-house e-mail system or changed the phone system. There always seems to be some people who never quite "get with the program." They have difficulties in adjustment and resist (maybe unknowingly) any change. Think of your route to work and the difficulty in getting to work when a situation such as construction or an accident causes a detour or delay. Accepting change can truly be a stressful situation for many people.

Attempting to change a safety culture can also cause stress at the workplace. But consider a starting point: where should one start? The answer is simple—everywhere! And the best place to initiate change everywhere is at the top. So, start at the top; the changes will work their way down to the bottom. Once the changes get to the bottom, the effect of these new-found changes and more will work their way back up to the top.

There are many reasons that change can be difficult to institute:

- As organizations mature, prosper, and diversify, management may be less willing to initiate any new ideas or changes. (Why not stay with the values that got us here?)

- Workers can become set in their ways and resistant to change. (This can occur with employees who have many years of service or new hires.)
- The person in charge is behind the program with lip service only (a nonbeliever).
- The talk is talked, but a considerable distance occurs between the talk and the walk.
- Past programs have been poorly planned out, poorly backed, and basically ended up not being implemented. This history makes it difficult for anyone to really get behind proposed change or support a new program.

If these pitfalls can be overcome, the process of building an effective safety culture can begin. It will be easily observable that workforces can and will change quickly when the chief executive is behind the proposed changes. So, the first thing that needs to happen is that the top management or the chief executive needs to believe that a safety culture change is needed. This top management person must have the vision to flourish rather than just survive.

Once the chief executive believes in the need for an improved safety culture, there must be a commitment of the necessary time and resources to initiate the changes. The time and effort needed to accomplish this task will likely be difficult, if not impossible, to accurately quantify. The expected gains, however, should be much easier to quantify. If you achieve zero incidents, your organization should have no accidents, losses, workers' compensation claims, equipment damage, or other losses. These are all concrete numbers that you can estimate, but raw numbers of losses are only a small percentage of what losses truly cost an organization.

In one of the last chapters of this book, I comment about a study that OSHA publicized on its website. This study contains information gathered during a variety of OSHA visits to different facilities. In one section of this study, OSHA asked workers what their perceptions were of the obstacles workers faced in performing their jobs in a safe and healthful manner. In another section of this same study, OSHA asked supervisors what they believed were the biggest obstacles they faced in performing their jobs in a safe and healthful manner. I think the answers from the workers and the supervisors will surprise you.

Before delving into assessment and improvement methods, the makeup of accidents is discussed. This discussion should shed some light on why accidents occur. After discussing accidents and why they occur, I present a management theory. In the draft stages of putting this book together, the next chapter was originally entitled "Modern Management

Theory and Philosophies." After considerable thought, the name was changed because I believe that it really discussed (Not So) Modern Management Theory. The theory cited throughout this book comes from a classic psychological study that took place in the 1920s and 1930s in the suburban Chicago area.

In fact, many of the basic concepts, theories, and philosophies presented in this book have been around for many years. Even though this is the case, I emphasize this classic theory because it appears that modern management often seems to complicate issues and disregard the basics. Keeping the fundamentals in mind along with a desire to keep things simple and easily communicable will help ensure that any programs will be successful.

This book focuses on the assessment and improvement or development of safety culture. After the (not so) modern theory discussion, it begins where one who is interested in changing current safety culture might begin—the assessment phase. Just like a physician might examine a patient while performing a basic physical, a similar examination should be performed on the safety culture. Once we determine where we are, we can determine the steps we need to take and the programs that need to be implemented so that the safety culture will develop and improve.

2
Management Philosophies and Theory

In this chapter, the causes of worker behavior are discussed. In theory, if we know why team members act in a certain way, we can work on and ideally change that behavior. This discussion is not about behavior-based safety systems but about basic worker behavior and how that behavior relates to safety performance. Safety statistics typically indicate that most accidents are caused by unsafe acts. Yet most companies have policies in place requiring that work be performed in a safe manner. These last two concepts appear to conflict.

On one hand, it appears that most accidents are caused by unsafe acts, but on the other hand, management requires and implements procedures so that unsafe acts are not committed. Something does not add up. Could it be that workers do not believe management? Or is it that management put those work safety policies in place but that they don't really mean it? Or could it be that workers enjoy hurting themselves? Maybe committing an unsafe act satisfies some kind of a basic urge to test fate. Maybe some workers enjoy self-inflicted pain. This chapter and the remainder of this book discuss not only why workers continue to commit unsafe acts, but also what can and should be done to eliminate accidents and unsafe acts in the workplace.

How management treats team members has an important effect on how these team members perform their job. We need to determine if the current company or corporate culture concentrates on integrity. If our management team is not "walking the talk" regarding safety aspects, there may be a tough road ahead for anyone who wants to implement change. Management's attitude toward workers greatly influences the culture. Much of this chapter is dedicated to demonstrating the relationship between management and team members, but for now we must realize that management needs to reach out to team members with openness and information sharing in an effort to improve. The general culture and safety culture in any organization should always be evolving and improving.

The supervisors and management leaders need to be groomed and trained so that they treat all team members as just that—team members. If we equate the different types of workers with parts of the body, the supervisors and middle management would be the backbone. Certainly, strong people with strong character and integrity need to fill these important positions. The team members and the people in the organization who actually get the product out the door are likened to the arms and legs. It is difficult to get anywhere or do anything without the arms and legs being in tune with the rest of the body (see Figure 2–1).

Top management is likened to the head or the brains. Vision, leadership, and support must come from the top of the organization. These aspects are important and are certainly necessary for a safety culture; however, not having all of the body parts working in unison handicaps

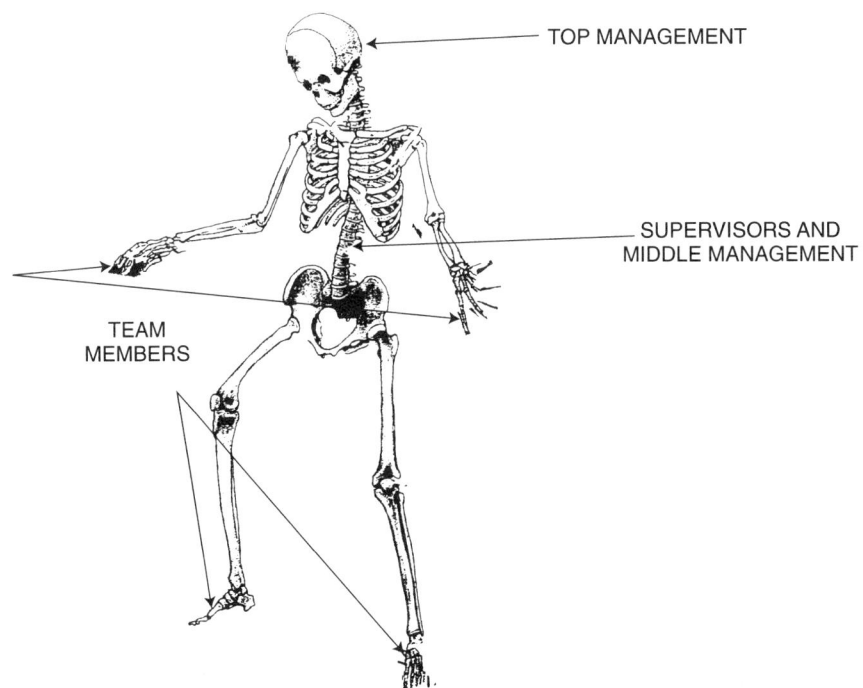

Figure 2–1 Parts of the Body vs. a Company's Makeup. A company's makeup can be compared to the parts of the human body. Although each body part has a different function, the body enjoys optimum performance when all of the parts are working in unison.

the organization. The job will probably still get done, but with certain segments out of unison, the other segments may have to strain to pick up the slack.

Some people would argue that accidents are an inseparable part of work. I believe that that idea is a myth. Here are a few more myths to add to the list:

- If there were no accidents, it is likely that no "real" work was performed.
- Accidents are just part of the cost of doing business.
- People are complex and unpredictable. No one really knows why people do what they do.
- Everyone is different; that's what makes the world go around.

I have been associated with many large projects where the projects experienced no lost-time accidents. I have also been associated with some projects that experienced no recordable injuries. When I examine the first two myths, I can verify that these points are not valid. Or at least that they do not *have* to be valid. Not to say that accidents are not part of doing business, but that they do *not* have to be.

I agree with the last two myths, but only to a certain extent. I believe that people tend to act in ways that will not injure themselves. I have to agree that people are complex and unpredictable, but I also have to say that accidents do not have to be included as part of the unpredictable or unsafe behavior. I believe that safety is a universal value that is shared—or at least should be shared—by all team members at all levels.

I have been asked throughout my career why people commit unsafe acts and become accident victims. The answer to this question is culture. I believe that the acceptance of workers committing unsafe acts is part of a poorly developed safety culture that needs to be changed. But before continuing the discussion on changing safety culture, let's analyze some accidents and see what conclusions we can draw.

ACCIDENTS ARE LIKE ICEBERGS

If your organization has experienced an accident in the recent past, there are probably many underlying factors and root causes behind the actual accident occurrence. Let's imagine our business as an ocean liner cruising along in the North Pacific. One of the pitfalls of a ship cruising the North Pacific is icebergs, just like one of the pitfalls of running a business is accidents. Let's analogize an accident to the iceberg. The accident can hurt our business the same way that the iceberg can hurt an ocean liner. They both poke holes in our program and potentially cause both the ocean liner and business to take on water.

A variety of analogies can be drawn between accidents and icebergs. If we were to witness an accident, that would be like seeing the tip of the iceberg. We can easily see the tip of the iceberg. So we can quantify it, count it, and attempt to steer clear of it. Once an accident occurs, we can do much the same thing: we can quantify it, count it, and attempt to stay clear of it in the future. But accidents—like icebergs—both share another dangerous trait. In order to avoid the iceberg, we need to realize that more than 90 percent of it is out of plain view. In order to avoid the accident, we need to realize that the true cause of the accident is likely not in plain view either. The true cause of an accident might include items such as the root cause, contributing factors, underlying factors, and a variety of other factors that are not easily quantified. So, if we want to steer clear of accidents, we need to realize that more than 90 percent of our solution is not going to be visible, or at least not in plain view (see Figure 2–2).

Even though it is difficult to quantify what the true cost of an accident really is, we can quantify the tip of the iceberg. And the tip of the iceberg for an injury-type accident is medical costs, workers' compensation, and litigation and award fees. A lot of research (with conflicting conclusions) has investigated the "true" cost of an accident. I believe that the cost is substantial and that—like an iceberg—90 percent of the cost of an accident is hidden or indirect.

Let's look for a moment at how labor costs are affected by an accident. Let's consider a case where a worker has an accident that causes two weeks of lost time and four weeks of modified duty. During this six-week period, the worker was not performing his job, so someone else had to do it. If this worker averaged 50 hours per week, the labor pool would have to have increased by 300 hours. In this age of being lean and mean, the 300 hours could be substantial because the replacement worker needs to be processed (i.e., applied and accepted, deemed fit for duty, hired, trained, oriented) before placement.

And getting a replacement worker may not be easy. A lot of other factors come into play, such as a slowdown in production because of using a replacement worker. A lot of other effects have been documented that stem from an accident, and none of them are good. This book will not delve into the many hidden costs of an accident occurrence in much detail, but I will use and refer to the iceberg and the 90-percent hidden costs as a reasonable estimate.

14 Maximizing Profitability with Safety Culture Development

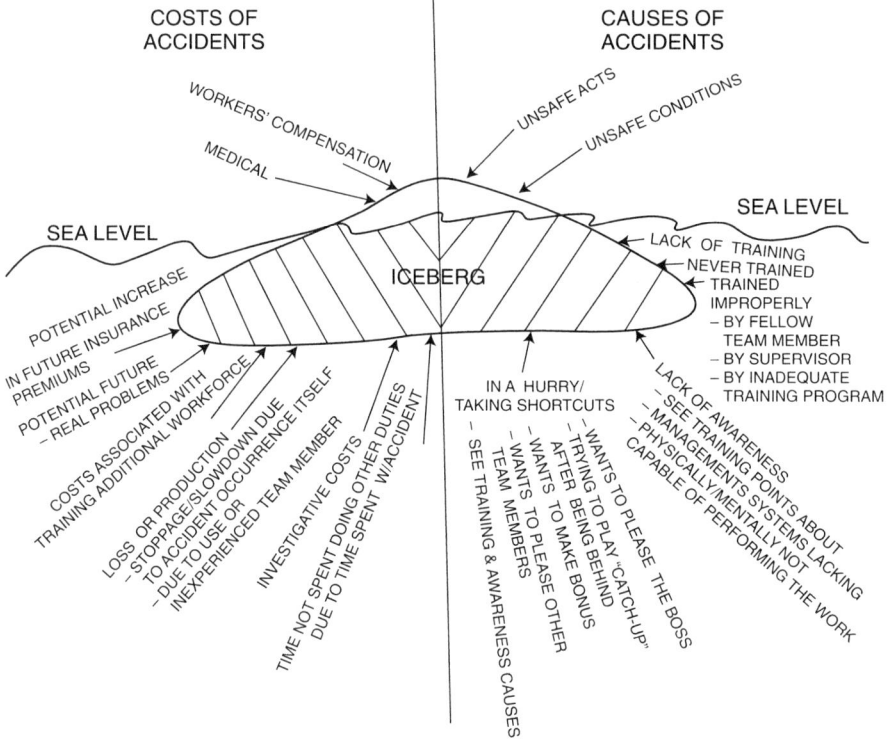

Figure 2–2 Accidents Are Like Icebergs.

(NOT SO) MODERN MANAGEMENT THEORIES

Management uses many modern theories and trendy techniques to motivate workers and attempt to prevent workers from performing unsafe acts. Some obviously work better than others, depending on the circumstances. Some theories are more complicated than others. Some have longer or fancier names. Whatever works best for your company should be the program you are using.

"If it ain't broke, don't fix it." "The right tool for the right job." Both of these sayings have been thought to be a sound philosophy, but there appears to be a flaw, at least in the first quote. If your company is experiencing work-related accidents, I think something is broken, and it better get fixed. As far as using the right tool for the right job, I have to agree; however, in the safety field, it can be difficult to determine what the right tool is.

During the course of my career in safety management, I have seen companies try to improve their safety culture in a variety of ways. Some of these approaches included the Deming Philosophy, the Safety Observer Program, Total Quality Management, and Behavior-Based Safety.

Under the right conditions, with all of the participants (especially top management) understanding the concepts and willing to do what it takes in time, effort, and general commitment, I believe that all of the modern-day management systems will work. In recent times, however, it appears that many of the systems are complicated, at least to the outside observer. Some of these systems seem to come with a whole staff of consultants to ensure that the program gets off the ground and heads quickly in the right direction.

I have nothing against consultants. Sometimes using consultants may be the best tool at a company's disposal to *really* effect change. The point is that many of these systems seem complicated and expensive, but they all seem to stress one underlying core belief, which is really quite simple. So instead of getting complicated, let's look at what I believe holds many of these systems together.

THE HAWTHORNE EFFECT

Have you ever heard of a study from which a basic management theory called the Hawthorne Effect was named? In this section, I summarize this study and the conclusions that were drawn from it. Even though this study took place almost 80 years ago, the conclusions remain valid today. This study gets into team members' basic needs. This theory still works because humans' basic needs have not really changed much in the last 80 years.

The Hawthorne Effect gets its name from a series of experimental field studies conducted between 1924 and 1932 at the Hawthorne Works of the Western Electric Company. The main product that the Hawthorne Works produced was the telephone. At this point in time, the telephone was a heavy piece of complicated wiring, relays, coils, and materials that were assembled and packaged in a labor-intensive manner.

The studies were designed to examine the effects of different working conditions, such as changes in rest cycles and illumination, on worker productivity (e.g., rates of assembling parts and winding coils). The surprising outcome from several of these studies was that each change in working conditions, even seemingly adverse changes such as decreasing normal workspace illumination by 50 to 70 percent, resulted in *increased* worker productivity.

Furthermore, at the conclusion of the first phase, productivity increased in response to changes in weekly and daily work schedules, lengths of rest periods, and the amount of lunch time provided by the company. Even more surprising was that production rates remained at an increased level even *after* workers were returned to prestudy work conditions.[1]

In November 1924, the Western Electric Company, in connection with the National Research Council of the National Academy of Sciences, planned to study the "relation of quality and quantity of illumination to efficiency in industry." In other words, the researchers were planning on changing the lighting in the work area to provide more or less light in different work areas, and then study any differences in production. This particular study lasted until April 1927, a period of two and one-half years.

Three different departments were selected for the purposes of the test. The procedure for each department was the same. First, there was a preliminary period during which the operatives worked under the existing lighting installation supplemented by daylight. Average production rates obtained during this preliminary period furnished baselines for calculating any future production changes. The level of supplemental illumination was then increased at four different levels. It was observed that the production levels corresponding to periods of different lighting intensities were always higher than the starting level and did not always fall off with a decrease in illumination.

It appears that the test results from this first phase of illumination testing somewhat baffled the testers. So, for the second illumination experiment, they divided one department in half. They named one half the "test" group and the other half the "control" group. They separated each group to minimize competition between the two groups. By doing this, the researchers running the experiment felt that the results they obtained would be directly related to differences in illumination intensity; however, note the following quote from Mr. Snow's report:

> This test resulted in very appreciable production increases in both groups and of almost identical magnitude. The difference in efficiency of the two groups was so small as to be less than the probable error of the values. Consequently, we were again unable to determine what definite part of the improvement in performance should be ascribed to improved illumination.[2]

The testers believed that lighting controls were indefinite because in the previous two experiments the illumination source was a combination of both natural and artificial light. So, a third experiment was conducted. In this experiment, only artificial light was used. The groups were divided as before into a test group and a control group. The control group was provided with a constant level of 10 foot-candles of light. The test group was provided with intensity levels from 10 to 3 foot-candles in steps decreasing 1 foot-candle at a time.

The results of this third experiment documented production increases in both groups; however, when the light level reached 3 foot-candles, Snow remarked, "the operatives protested, saying that they were hardly able to see what they were doing, and the production rate decreased."

Mr. Snow then placed a work station in a darkened locker room and had operatives perform their work using the equivalent of moonlight. Even under these very dark conditions, the production levels remained high.

The testers feared that the effects they were recording were more a result of psychological reasons than illumination, so they modified the experiment further. They took a test work group and work area and increased the lighting at regular intervals every few days. After a while, electricians were asked to change lightbulbs, but not increase the bulb size or output, as they had been doing in the past. The workers were told that the lighting was being adjusted upward, and the work group in the area reported that they preferred the increased light. (Keep in mind that there were no increases in light at this time.) After a few days of this, the tester had the light decreased each day, and the workers continued to report that they preferred increased light. Interestingly enough, whether the light was increased, decreased, or kept the same, production stayed at essentially the same high level.[3]

In the conclusion of Chapter 1 regarding this study, the testers felt that the experiment "fell short of the expectations." The reasoning for the shortcoming was because of the failure to show the relationship between lighting and efficiency.[4] Later the focus of this experiment shifted from light adjustments toward break and rest periods.

The testers believed that the general improvement in rate of output had been caused by the introduction of rest periods; however, no matter how the rest periods were adjusted, production increased steadily. This trend of increased production confounded the testers and caused them to look further for the reasons production increased. From the report, "it became necessary to seek a new hypothesis to explain certain unexpected results of the inquiry."[5]

By October 1929, 60 percent of the workforce was enjoying a morning and afternoon break in addition to the lunch period. Before this study, only a lunch period was considered a formal break or rest period. As the additional breaks were instituted, the Hawthorne Works recorded increased production in each department instituting these additional break periods.

What impressed management most, however, were the scores of latent energy and productive cooperation, which clearly could be obtained from its working force under the right conditions. Among the factors creating these conditions, the attitudes of the employees stand out as being of predominant importance. All attempt to relate the experimentally induced changes to the operators' performance, apart from their effect on the operators' attitudes, had been inconclusive. Management decided, therefore, that everything pointed to the need for more research on employer attitudes and the factors to which they could be related.[6]

X, Y, Z WORKERS

Let's discuss a little bit of theory. If you reach back and think of basic worker theories, you might remember X-, Y-, and Z-type workers. In 1960 Douglas McGregor published a book called *The Human Side of Enterprise*. This book discussed theories X and Y at great length. One philosophy was that workers were basically evil, lazy, and not willing to be "team members" because their basic nature will not allow them to do so. This is called theory X. Theory X carried with it the following three main assumptions:

1. The average human being has an inherent dislike of work and will avoid it if he can.
2. Because of this human characteristic of dislike of work, most people must be coerced, controlled, directed, or threatened with punishment to get them to put forth adequate effort toward the achievement of organizational objectives.
3. The average human being prefers to be directed, wishes to avoid responsibility, has relatively little ambition, and wants security above all.[7]

Supervisors who practice or believe that team members basically follow theory X behavior will coerce, threaten, punish, and so on in order to motivate workers. The theory X supervisor believes that workers are basically "lazy" and "bad." If the supervisor is not *on* the workers at all times, the workers will always take advantage of the supervisor. Theory X supervisors do not trust the workers they supervise. In today's terms, the theory X supervisor might be called a micromanager, and probably a lot of other terms we won't print here.

Besides theory X, we have theory Y. Theory Y has the following assumptions:

1. The expenditure of physical and mental effort in work is as natural as play or rest. The average human being does not inherently dislike work. Depending on controllable conditions, work may be a source of satisfaction (and will be voluntarily performed) or a source of punishment (and will be avoided if possible).
2. External control and the threat of punishment are not the only means for bringing about effort toward organizational objectives. Man will exercise self-direction and self-control in the service of objectives to which he is committed.
3. Commitment to objectives is a function of the rewards associated with their achievement. The most significant of such rewards (e.g., the satisfaction of ego and self-actualization needs) can be direct products of effort directed toward organizational objectives.
4. The average human being learns, under proper conditions, not only to accept but also to seek responsibility. Avoidance of responsibility, lack of ambition, and emphasis on security are generally consequences of experience, not inherent human characteristics.
5. The capacity to exercise a relatively high degree of imagination, ingenuity, and creativity in the solution of organizational problems is widely, not narrowly, distributed in the population.
6. Under the conditions of modern industrial life, the intellectual potentialities of the average human being are only partially utilized.

Theory Y is another basic philosophy, but this one theorizes that workers are basically "good," or at least that workers are not inherently "bad" or "lazy." Theory Y implies that if workers are lazy, indifferent, unwilling to take responsibility, uncooperative, and so on, the responsibility for this behavior lies in management's methods of organization and control.

Workers will basically try to do a good job when they are properly managed. A supervisor's (or any manager's) role is to guide, coach, and counsel workers, with discipline being a last resort that is seldom used. The emphasis by management at all levels must concentrate on the positive. There may be times when it appears that workers are not responding to positivity or will resist continuous performance improvement, but, for the most part, management must continue positive reinforcement, especially for safety-related items.

So, if we want our businesses to continually improve, management needs to find ways to motivate team members and keep their awareness levels high. The Hawthorne Effect is a basic motivational theory that has been shown to work. If management treats people as an important part of the team, workers will respond in a positive way. The Hawthorne study showed that if supervisors were more helpful observers and coaches, workers would respond more favorably than if supervisors were disciplinarians. Having workers feel responsible for the quality of their work, and not that quality is the quality control department's problem, was noted in the Hawthorne study.

In the 1950s Japan was attempting to rebuild its economy post-World War II. Japan had undertaken a massive effort to export goods, but had been unsuccessful in its rebuilding. Japanese goods were felt by many people to be "cheaply" made or just plain "inferior" products. And in some cases, this inferiority or "cheapness" appeared to be a concern. As we all know, the Japanese economy did become very powerful and no longer is associated with the production of inferior goods. In fact, in many areas, including electronics and computers, the Japanese are world leaders and known for producing the finest of quality of goods. The Japanese attribute much of their success in manufacture to total quality management (TQM) techniques to an American named W. Edwards Deming.

In fact, the Japanese named their top-quality prize after Deming, who in the 1950s taught them many of these methods. Deming is known as the father of TQM theory. Deming believed that management shortcomings and not "bad" workers caused most product defects. Deming believed that getting it right the first time was much more important than inspecting the work at the end of the production line.[8]

Deming picks up this important point as one of his key points in successful management noted in the 1960s and 1970s. Keep in mind that one of Deming's main points is that workers should be their own bosses or supervisors. Supervision should play the role of a facilitator more than a disciplinarian. Thinking along similar lines, safety shortcomings should not be the job of the safety department.

The programs discussed in this book stress what I believe to be the best management practices. These practices have been field tested and proven to work. The book stresses the Hawthorne Effect throughout because I believe that the Hawthorne Effect is the basic reason that workers act the way they do.

Simply stated, if management can create an atmosphere in which workers are treated as part of the team and are given attention and each team member is a valuable asset to the company, workers will be motivated and empowered. In regard to safety, a motivated worker will be attentive, and an attentive worker is a safe worker.

On the other hand, if workers feel that they are not team members, that they are not appreciated, and that the work they do has little importance, you will likely find an inattentive worker who has developed a poor attitude and will likely soon be a safety problem.

Example: Unsafe Widget Stacking

Let's say that a worker is performing a certain work task outside of the acceptable safety practices. This work task could take on any form, and just for this example let's say this worker is stacking widgets. Stacking widgets in this unsafe manner has been "business as usual" for years, and all of the workers in this department have recognized that the practice is unsafe and directly conflicts with the company policy. This practice continues with management well aware of what is going on. Management walks by regularly, has observed the unsafe act, but looks the other way. The unsafe act goes unaddressed.

This situation is typical in the daily activities of modern business. Unsafe acts continue to be the culprit cited as most responsible for accidents. Management continues to believe that they are doing everything they can to eliminate accidents. And workers continue to believe that they are caught in the middle with wanting to please management by getting the job done "by the book," properly, quickly, safety, and so on, and, not compromising any company policies, especially safety rules to get there. Some workers believe that if they want to get their job done, at the same time, they *have* to take shortcuts. And taking shortcuts with safety is similar to taking shortcuts in quality or production.

Shortcuts that compromise safety, quality, production, and so on just "don't cut it" and should not be allowed under any circumstances. Instead, doing things right the first time must be stressed. Our business should strive to be better, safer, cheaper, and faster than our competitors. Continuous performance improvement programs should be instituted that include safety as an integral part of the business. Our standards should be set high and adhered to.

When examining the unsafe widget stacking example, a couple of key points come to mind. If management has allowed the unsafe widget stacking to continue without taking corrective action, they have (maybe unwillingly) given the signal that it is acceptable to stack widgets in an unsafe manner. Management has likely made a bad situation worse by giving the worker(s) recognition regarding the job they have been doing. In fact, if this situation goes on for a period of time, it is likely that the employee who is performing the unsafe task has had some performance reviews. If nothing is mentioned during the reviews regarding unsafe work habits, management is encouraging this unsafe act to continue. If nothing is done to stop this act, management is providing a form of positive reinforcement that sends workers the signal that it is okay to do things the wrong way.

That is, workers realize that they are not supposed to stack the widgets in an unsafe manner, but for whatever reason (we will discuss the reason later), the worker continues to choose the unsafe manner. Management knows of the shortcoming but chooses to ignore it. The worker continues to perform this task while the paychecks and rewards keep flowing. Therefore, the worker gets encouragement (positive reinforcement) to perform unsafe widget stacking.

If management tells a worker one thing (i.e., stack widgets in a certain safe manner) but fails to follow through (ignores the situation) we have the makings of an unfortunate situation. This feeling that management does not practice what they preach can have far-reaching poor effects on our safety culture. Relating this back to the Hawthorne Effect, does the team member whose unsafe act gets ignored by management feel like a valuable team member? I think not. If we look at the Hawthorne study, we should be giving all of our workers adequate attention and guidance so that they know they are valuable team members.

WHY WORKERS PERFORM UNSAFE ACTS

There are likely a variety of complex reasons why workers perform unsafe acts. If we continue to discuss unsafe widget stacking, there might be a lot of reasons why the worker is performing in this unsafe manner. Maybe there is a lack of training. The person who trained him how to do this was also doing it in an unsafe manner, and the unsafe way of doing things just got passed down. Maybe nobody trained this worker.

The need to put an untrained worker into a position can arise for a variety of reasons (e.g., absence, tardiness, vacation). The trained person is not available, and the job needs to get done. Someone needs to perform this task. Maybe the supervisor assumed that the fill-in person knew how to do the job safely. Or possibly the supervisor or training department could not schedule the time to adequately train the fill-in person. After all, the supervisor might have learned his job the hard way—sink or swim. Some people still believe that getting thrown into water over your head will make you into a good swimmer quickly. I prefer training and getting used to shallow water with a slow progression into deeper water as the trainee shows that he or she can handle it.

When someone is performing work tasks in an unsafe manner, we need to delve deeply into the situation. When an unsafe act is observed, management should ask the question, "Why?" Why is someone doing something that they are not supposed to do? When you arrive at the first answer, ask the question "Why?" again. And keep asking the question "Why?" until you have arrived at the core factors.

Once you have discovered the core factors, examine each factor and determine what type of corrective action should be undertaken to eliminate that factor. Assign a person to be responsible and a target date for each factor. Reexamine the situation regularly. Reassess the situation and make new recommendations with new responsible persons and target dates until the situation has been resolved. The resolution of these situations often will not be easy or quick. Some will be struggled with for months or years. When a difficult situation does get resolved, the positive effect on safety culture will likely be tremendous.

Example: The Slippery Slope of Safety

> Many years ago, while working as a safety manager, I was responsible for safety in a water treatment plant. Treatment plant operators were required to climb ladders to the tops of million-gallon switch-batch reactor tanks and then walk along the top of the tank to the middle, where a hatch would be opened and the tank could be inspected. The tanks were "crowned" on the top, which made them difficult to walk on. This difficulty became even more challenging when changes in the weather or humidity made the tops of these tanks slippery.
>
> The situation got attention when the operators voiced a concern during a safety inspection. After some discussion, the operators were given harnesses and lanyards and asked to tie off when they reached the top of the tank. The safety committee felt that a long-term solution would be an engineered guardrail system around the tops of all the tanks that would include a toprail, midrail, and toeplate. This system would eliminate the need for a tieoff and improve the overall safety of the job.
>
> *(continued)*

This all sounded good, and each month during the safety meeting, the situation was reviewed. It turned out that the cost of these rail systems was determined to be substantially above what anyone on the committee had estimated it to be. This cost was *way* above what the local manager could order without approval. The committee was asked to come up with other alternatives. Other alternatives were discussed over the next two months, but none of the alternatives that were discussed seemed to provide an adequate solution. The consensus was that the engineered guardrails should be pursued. With a little red tape, and a couple of months' wait, the guardrail systems were ordered.

However, these guardrails were not an in-stock item; they needed to be fabricated. The fabrication and delivery of these items took another three months. When these items finally arrived, we discovered that the vendor had sent the wrong pieces. With a little urging and some rush shipments, the improper pieces were shipped back and proper pieces arrived. These guardrail systems were installed on all of the switch-batch reactor tanks within the next couple of months.

Although this may not seem like a large accomplishment, it gave me great personal pleasure to take this item off the list. I was not sure what effect this ordeal had on the safety culture of this plant. I personally felt gratified and relieved that it was done, but was disappointed that it took so long. Shortly after this guardrail system was installed, my responsibilities shifted and I was no longer responsible for safety at this plant, and life went on.

Approximately three years later, I was touring a large site in Ohio (which was located about 500 miles from the aforementioned switch-batch reactor plant). During my tour, a worker whom I hardly recognized came up to me. He was one of the operators of the switch-batch reactor plant. After reintroducing ourselves, the first words out of this operator's mouth were to thank me for the persistence I exhibited during the difficulty with the guardrail systems. I remembered that this operator was a bit quieter than the others during this process, and I had often wondered how he felt about it. His comments were that the guardrails had saved him from falling numerous times and that he appreciated them most when the tops were slick and he was working the midnight shift alone. In retrospect, he was not disappointed in the time that the guardrail system took to complete.

Now, let's get back to the subject at hand, which is why do workers commit unsafe acts? After we have asked ourselves "Why?" several times and found core reasons for shortcomings, let's ask ourselves a couple more questions. The first question I would ask would be: "Have we performed a job safety analysis (JSA) on this task?" If the answer is "no," we need to do that right away. If the answer is "yes," we need to revisit the JSA and see where that leads us. If we are not following the guidelines in the JSA, we need to determine why we are not following them. The JSA is discussed in detail in Chapter 7, "Training."

When used properly, the JSA is a valuable tool in the development of culture, as a training aid, and to emphasize the Hawthorne Effect. JSAs should be an important part of any company's safety culture and are referred to in several areas of this book. The Occupational Safety and Health Administration (OSHA) has published extensive information in a pamphlet available in hard copy and through the Internet from the OSHA web page. For the reader's convenience, I have included a copy of the majority of this document as Appendix A. Keep in mind that although the information found in Appendix A can be useful, the format offered in Chapter 7 has the advantage of being easily understood by the team member performing the work task.

THE NEED FOR CULTURE: SELLING THE PROGRAM

How do you know that a culture gap exists? One main indicator is accident occurrences. If the company you work for has accidents and accepts them, it is in need of a safety culture improvement. If the company you work for is in need of safety culture improvement, then selling this program, or at least getting started, should be a relatively easy task. How do we sell the need for an improvement of safety culture? One effective method includes first convincing those at the top of the organizational chart. One convincing argument will include cost savings.

The safety department is sometimes considered a drain on a firm's bottom line. I believe that just the opposite is true. The safety department should be looked at as a profit center because it can save the company many thousands of dollars when operating effectively. Let's assume that our widget distributor department grossed $1 million last year and made a 10 percent profit. Let's say that the average accident in the widget department was $10,000 in workers' compensation and medical costs. If our 90 percent iceberg assumption is accurate, then this accident really cost our firm $100,000. This means that without the accident occurring, they could have doubled the profit in the widget department last year.

Even if we can't sell the 90 percent iceberg theory, let's just look at the direct cost of $10,000. This firm would have to sell another $100,000 in widgets just to pay for the one average accident—that's a lot of widgets. If we consider the iceberg (or even if we don't), one should be able to make a powerful case for safety culture improvement. Preventing one accident could prove to be a huge savings and increased profits for any company.

Selling safety culture improvement is not delved into too deeply in this book. Volumes have already been written on how to effectively sell a program. The subject is vast, and this book only touches the surface of selling the program. Instead, this book primarily focuses on the development of safety culture and safety programs.

SUMMARY

In the first two chapters, we have gained insight into what appears to be a conflict. Management implements systems so that workers will not get injured; however, even though these systems are put in place, workers still get injured. An analogy has been drawn between accidents and icebergs. Basic worker behavior has been discussed, along with management's attitude about workers. Another analogy has been drawn between a team member's role in the workplace and parts of the human body.

The Hawthorne Effect demonstrated that workers will be more productive when they are treated like valuable team members. As the management at the Hawthorne Works plant demonstrated that workers were important, the production in the plant increased.

Supervisory styles were mentioned both in the Hawthorne study and in a book discussing theory X and Y. The theory X manager/supervisor thinks that workers are basically bad and cannot be trusted. Punishment is mentioned as a motivator for the theory X supervisor. The theory Y manager/supervisor believes that workers are not basically bad. This theory attributes poor worker performance to a management failure and not to the worker's nature. Later chapters expound on the basic themes presented so far and on developing programs to enhance safety culture.

REFERENCES

1. Borgatta, Edgar F., and Marie L. Borgatta. *Encyclopedia of Sociology* (New York: Macmillan, 1992), 793.

2. Spurge, Lorraine, ed. *Business Encyclopedia* (Knowledge Exchange, 1997), 548–49.

3. Ibid.

4. Ibid.

5. Ibid.

6. Ibid.

7. Roethlisberger, F. J., and W. J. Dickson, *Management and the Worker* (New York: John Wiley, 1964), 14–19, 59, 185–86.

8. Spurge, *Business Encyclopedia*.

9. McGregor, Douglas. *The Human Side of Enterprise* (New York: McGraw-Hill, 1985/1960), 33–5, 47–8.

3
Assessment Tools

The occurrence and acceptance of work-related accidents is a reflection of the development of a company's safety culture. This chapter concentrates on accident trends. Because accident trends can be elusive, this chapter also offers some ideas for how to ensure that the reviewer is getting a "true" picture of the accident occurrences.

Besides the use of accident trends, other tools can be used to effectively assess culture. A short discussion of some useful tools to determine safety culture development is offered later in this chapter. Later chapters delve into a selected number of these tools in much greater detail.

ACCIDENT TRENDS

How does one determine accident trends? The answer to this question may not be as straightforward as one might think; however, looking over the various tools in the toolbox, we can begin to get a good view at what has happened in the past. Getting a good feel for the types of accidents that have occurred in the past will tell us a bit about the development of the safety culture.

Determining the development of your company's current safety culture is a top priority because it is difficult to set a course for the future if you do not know your current location. So a logical first step is to examine accident occurrences in the near and recent past and determine trends. A lot of tools are at our disposal to aid in making this determination, including the following:

- OSHA logs (such as 200 and 300 logs)
- First-aid/incident logs
- Individual accident reports
- Accident investigations
- Medical records

OSHA Logs

One obvious place to start is your OSHA 300 log. By law, this log needs to be fully completed and accurate. If we are going to be law-abiding citizens, this OSHA log should be thought of as a useful tool and used as such; however, even if the company management is law abiding, the OSHA log often does not always tell the whole story. Note the example that follows.

Example: OSHA Log Compliance

I was reviewing past OSHA logs (they were called 200 logs before 2002) in order to determine if what had been entered on these logs should, in fact, be recorded. I gathered the medical file for every case on the logs that I was asked to examine. Most of these cases had been entered years before I had joined this particular company. Many of the people mentioned in these reports had retired or moved on, so I was left with only the accident reports and medical records to tell the stories. While looking through the medical files, I was able to legitimately remove several cases that were first aid rather than OSHA recordable.

One case that stands out in my memory was that of a worker who had gotten a metal sliver in his eye and went to the clinic for removal. The person responsible for making the recordability call immediately recorded this injury because of what he interpreted to be the currently in-place OSHA guidelines. The guidelines at this point in time about removing objects from the eye were straightforward. That is, when forceps, tweezers, or other mechanical devices were used to remove material from a worker's eye, the incident must be considered OSHA recordable; however, when I closely examined the medical records, the report stated that the metal sliver was in the eyelid and not the eye.

(continued)

> My decision to remove this incident from the OSHA record logs was also straightforward. After all, there was no prescription medicine, no restriction, no change in job or loss of work, and therefore this case was removed from the OSHA record log.

Although OSHA recordability is an important point, we should not lose sight of the big picture. That is: what caused this accident? What are the reasons that this injury occurred? We need to determine the reasons for injuries so that we can prevent the recurrence. I am going to repeat this statement because the point is paramount: *We need to determine the reasons for injuries so that we can prevent the recurrence.*

Let's keep this very important point in mind, but let's also remember a couple of other very important points. The OSHA log is typically used not just for recording and tabulating injuries, but will likely be closely scrutinized if your company performs any subcontracting work and is interested in bidding these jobs. Clients want to know how safe any contractor is not just for the humanitarian reason, but also for the economics.

If your company has chosen to take the conservative stance on OSHA recordability of accidents, it will probably be perceived as less safe than a competitor that has a less conservative approach to OSHA recordability. The long and the short story is: *Your company may needlessly be losing awards because of a conservative stance on OSHA recordability.*

If your company is losing awards, it is probably also losing revenues, profits, and is missing a variety of financial goals. What can be recommended? The answer is simple: Follow the guidelines, but don't be extreme. If you think you are being extreme, get some other viewpoints. The current standard offers lots of examples and guidance regarding proper classification of injuries, but no matter how many examples are offered, there may be instances where the case you are considering seems not to match the examples offered.

Besides reading the standard, keep in mind that OSHA will typically have compliance officers give free presentations at functions at various times throughout the year. Typically, the safety professional at your company will likely be a member of an organization that meets regularly and discusses important safety issues like OSHA recordability. Over the past year, I have personally attended three different functions where the subject being discussed was OSHA recordability. Attending these functions provided an easy and effective way to speak face to face with a compliance officer and have questions answered or points clarified in a professional manner without putting anyone in jeopardy. I would suggest that if your company's safety representative is not an active member in the local safety association that he or she become one. Staying active in these types of associations is an excellent way to stay current and pick up new approaches to the performance of safe work.

Pursuing membership in the STAR program is another way to form a working relationship with OSHA. The company you are working for should ideally be on good terms with OSHA; however, unless your company is on the STAR Program or other similar status, it may be unlikely that anyone at your company deals with OSHA regularly. In my experience as a safety professional, I have dealt with various STAR facilities over the past several years. At each STAR facility, any injury/incident receives scrutiny by OSHA, along with the company representatives. The status of recordability of an incident even at a STAR facility is typically a matter of debate. Be as thorough as possible so that your decision on recordability reflects all available information. This would include everything from a thorough accident investigation, including statements from the injured worker, the supervisor, any witnesses, and the medical treatment facility.

If you have gathered all pertinent information and are still not sure of the status of the incident, you can always call your local OSHA office. Some have tried this approach both anonymously and not. Calling OSHA may still leave you with doubts. Although compliance officers are highly trained and knowledgeable in most safety areas, recordability may or may not be their specialty. In addition, the description over the phone may not give the full picture or provide full information to the compliance officer. Calling OSHA for recordability advice is similar to calling the IRS for tax advice. Although sometimes the calls do provide useful information, I would prefer to have something in writing.

If you are not in a big hurry, you may wish to state your question in writing in the most objective manner that you can and then submit your question to OSHA in the form of a letter. Simply state your case as clearly as possible and ask OSHA to reply in writing as soon as possible. As mentioned previously, being in a big hurry does not bode well for this solution. Although OSHA will eventually answer your letter, many months may go by before this answer is forthcoming. Also, remember that your letter becomes part of the public record and may be included in the database of interpretive guidelines along with your name and the company for which you work (if it is mentioned in the letter).

What I believe to be a better (and much quicker) alternative is to conduct a search of OSHA's current reply letters. Questions very similar to yours may have already been answered by OSHA and become official protocol. You can search their database for explanations on a huge variety of subjects.

For cases involving a potential OSHA recordability issue when the circumstances might be unclear, I would suggest putting a note in the medical file stating the logic of why this incident was recorded (or not recorded). If your records are audited by OSHA at a later date, at least you will have a written record of why the choice was made and a logical reason for making the choice. Issues of recordability can come up more often than you think.

Some safety professionals cannot separate workers' compensation compensability from OSHA recordability. These are two differrent issues guided by different rules. Just because an industrial commission rules that an injury is compensable does not mean that the case is OSHA recordable.

Example: The Monday Morning Blues

> I have been involved in numerous accident cases that have been termed the classic "Monday morning" injury. In this classic case, the worker is injured over the weekend, but shows up Monday morning and claims that the injury happened on the job. Sometimes these Monday morning injuries cannot be shown to be false and become compensable under the workers' compensation laws.
>
> If the case holds what I believe to be reasonable proof that a reported work-related injury is, in fact, not work-related (i.e., a Monday morning injury), I would *not* include the injury as an OSHA-recordable injury. I would likely *not* record this injury even if the industrial commission determined this injury to be compensable.
>
> I would carefully document this reasonable proof in the file. Reasonable proof might consist of witness statements, phone calls from spouses or outside sources, hearsay evidence that I believed to be true but that was not allowed in legal proceedings, and so on. Keep in mind that I mention *reasonable* proof. Let's not get carried away and start calling injuries that are claimed to have happened on Monday morning false reporting simply because they are reported on Monday morning. But, if there is reasonable proof that this occurrence is a "Monday morning" classic case, I would document my proof and likely keep this case off the OSHA log.
>
> On the other hand, I have obtained on several occasions useful information from workers' compensation proceedings that helped

me determine that a case should not be considered recordable. For instance, I was involved in an industrial injury claim for a back injury case that was assigned to a workers' compensation physician. The injured worker was back to work without restriction for some time when the Compensation Board ruled that the injury was not work-related. Although the injured worker claimed that the injury was work-related, and I had initially recorded this case as such, after reviewing all of the evidence I felt confident that this case could be removed from the OSHA log. Remember, all of the information needs to be weighed before making the final decision on recordability.

Accident trends are important when assessing a safety culture. We need to determine where we are now so that we can plan our moves forward. In order to help determine where safety culture has been and where the culture is at this point in time, the typical starting point and what some people consider to be a primary tool is used—the accident log. This log is like a road map and should be examined carefully. OSHA logs must be filled out properly using current guidelines and acceptable standard industry practices. Being too conservative can cause your business to potentially lose awards. Being too liberal in your interpretation may be considered noncompliant.

Unfortunately, the person who has been responsible for filling out the OSHA log probably used his or her best professional judgment when doing so. Anytime judgment comes into the picture, room for error is also a factor. Each accident case on the log is typically placed there because the person ultimately in charge of deciding what cases belong on the log felt that the merits of each case fell within the OSHA guidelines for recordability. OSHA sets up these rules to be applied universally so that all companies rate their accident occurrences in the same manner. Under ideal circumstances, if a team member at the widget manufacturer got into an accident that caused a certain type of injury that caused management to call this a recordable case, then when a team member is involved in the same type of accident at the competitor's site, it would also be called a recordable case.

In principle this all sounds good and fair, but in practice, it does not always work out this way. I have worked in companies that treated recordability with tremendous variability from division to division. The work that was performed at each location was essentially the same. One of the people in charge of determining recordability would record everything.

34 Maximizing Profitability with Safety Culture Development

This meant that anytime a person received medical treatment, no matter how minor, this case would go on the OSHA log.

On the other hand, the other division looked at accidents in exactly the opposite manner. Even if the accident appeared to match recordability, attempts would be made to exclude it from the log. The differences here come from within the same company. The logic that some people use is that "I am going to record everything. If I record everything, there is no way that OSHA will ever be able to say that I have falsified records or tried to hide anything." The logic going to the other extreme seems to be, "I am going to record as little as possible. After all, it is the employer's judgment to determine whether an accident is 'serious' or not."

Do you remember the Three Bears—not too hot, not too cold, and just right? The Three Bears method holds true for recordability. Recordability is a balancing act (see Figure 3–1). Each case is different and each injury is unique. The principles behind recordability are meant to provide guidelines for separating serious injuries from nonserious injuries. Certain barometers are used as part of the definition of serious injury. Keep in mind that the term *serious injury* is in the eyes of OSHA and not in my eyes or the eyes of the medical professional.

Over the years, I have had many conversations with medical professionals who interpret OSHA recordability differently than the guidelines

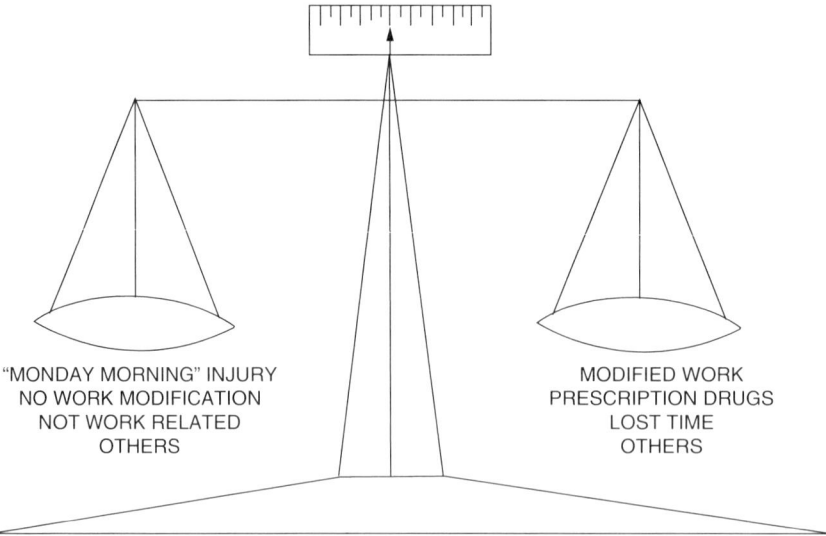

Figure 3–1 Balancing Recordability. OSHA recordability can have a serious effect on your company. All information regarding a case needs to be carefully weighed before making the final decision on OSHA recordability.

offered in the standard. Let's take a typical example—prescription drugs. I have had several conversations with physicians in various parts of the country who have advised me that a particular case should not be considered OSHA recordable even though prescription drugs were prescribed and administered for a week or more. The OSHA standard is clear on this subject. Although doctors use their judgment in the issuance of prescription medication, the employer and not the doctor must make the judgment of recordability. So, even if the doctor states that an injury should not be considered recordable, this opinion should have little bearing on the actual recordability of the case.

This book is not attempting to go into detail on how to determine OSHA recordability. Whole books have been written on this subject. Instead, let me suggest that you refer to the OSHA standard for definitive guidelines.

First-Aid/Incident Logs

Companies with a well-developed safety culture will likely have first-aid logs and incident logs. Getting people to fill out these forms when no injury was sustained or damage involved can be challenging. Most team members will resist this "extra" paperwork. People believe that fault finding and punishment of the guilty are the likely outcomes of this type of reporting. The fears that people have in reporting, especially when no injury is sustained, need to be kept in mind.

Recall from Chapter 2 the discussion on theory X and Y. There are still many supervisors employed today who are the typical theory X supervisor or manager. Caution must be exercised to ensure that these types of reports are used for accident-prevention purposes and not fault finding. Getting people used to filling out these reports takes time and determination. The format used in these report forms should be worded so that *prevention* is the key word.

The format used to report first-aid cases or incidents can be similar to an accident report or can be a separate form. I recommend that your organization develop the form for these types of incidents in-house. Getting input from the people who will be filling out these forms for their development will likely increase the likelihood that they actually get filled out. The first-aid logs can resemble the OSHA 300 log. Any injury that resulted in a Band-aid, ice, a wrap, or medical treatment that is not considered to be recordable would be included on the first-aid log. The name or other identifier of the victim would be included, along with a brief description of the accident, treatment, status, and so on.

The first-aid log is another important tool used in accident prevention. This type of log is not a requirement for OSHA. It is an in-house tool that, when studied carefully and acted on, will help reduce or eliminate accidents. If we believe in the pyramid effect, when we reduce the size of the bottom of the pyramid, which is reducing common minor incidents like first-aid incidents, this reduces the number of serious injuries.

An incident log is another useful tool for accident prevention. These are sometimes referred to as "near misses" or "near hits." I prefer the term "close call." An incident log may also be used to record failures or losses that did not result in an injury but could have.

Let me offer some examples of what would be included on an incident log. Let's say a crane dropped its load. The falling load did not result in injury to anyone, and the material that was dropped was not damaged. Even though no injury occurred, this accident could be referred to as a "close call" and should be included on the incident log. An incident investigation would take place, and reasons for the failure and ways to prevent a recurrence would be forthcoming and implemented.

Another example might be when a piece of machinery fails because a drivegear disintegrates, with pieces flying across the shop floor. No one is struck by the flying pieces, but if a piece of the gear would have struck nearby team members, serious injury would have been likely. As in the case of the dropped load, even though no injury occurred, this accident could be referred to as a "close call" and would likely be included on the incident log. An incident investigation would take place, and reasons for the failure and ways to prevent a recurrence would be forthcoming and implemented.

Close calls and nonpersonal injury incidents are important. They should be given the same attention or more than those incidents resulting in injury. As your safety culture develops, the more common the use of first-aid/incident logs as accident prevention tools will become.

Individual Accident Reports

Individual accident reports, sometimes referred to as Form 45s, employee accidents reports, or supervisor accident reports, should be filled out for every accident. Each state typically has workers' compensation guidelines that require reporting any work-related injury within a certain time frame. If your organization has not been keeping these records, you can expect a variety of things to happen. First, you will probably get bills from hospital emergency rooms for treatment for work-related injuries for which you have no record. Another likelihood is that your company will receive letters from lawyers demanding payments in regard to work-related accidents. Also, your state Workers' Compensation Board will likely be sending your company notices of noncompliance.

Your company must ensure that all injured workers report their injuries immediately (or within acceptable time frames) and that persons such as front-line supervisors ensure that documentation of the incident is submitted in the proper time frame. If and when work-related accidents occur (and I hope that your safety culture is so well developed that you have zero accidents), prompt reporting and documentation must be required.

The person(s) who is attempting to determine safety culture development should review all of the individual accident report forms. During this review, any shortcomings or needs for improvement should be noted. For instance, if these reports are incomplete or filed late, a determination needs to be made about why this has happened. The result could include policy adjustments or other corrective action.

Each blank space on the accident form that is applicable to the incident should contain the appropriate information. This will always include the date of occurrence, time of occurrence, worker identifiers, an account of what happened, and the name of the supervisor or person filling out the report. If any of this information is missing, the report should be returned and should not be accepted until the deficiencies are corrected. If your organization has accepted accident reports that are deficient, you need to address this situation with vigor. Keep in mind that the Hawthorne study needs to be your driver. The person responsible for filling out these reports needs to realize that these reports are *important*, and that management *cares* about these forms and cares that they are filled out within company guidelines and acceptable parameters.

Inadequacies noted on accident forms are an indication that the safety culture is in need of further development. Although we are aiming toward zero incidents, if an incident occurs, we must act swiftly and decisively. This includes the proper filing of forms or documentation.

Accident Investigations

Besides accident forms, accident investigations are a tool used to assess safety culture development. An investigation should be part of every accident, incident, or close call file. This investigation can vary tremendously in size and depth depending on the incident and its potential impact, but nonetheless an investigation should follow every incident.

If your firm does not investigate each incident, arrangements to change this situation should be made. This is an indicator that a safety culture gap exists, which needs to be filled. We will discuss accident investigations in Chapter 11, "Safety Culture Barometers."

Medical Records

When someone has sought medical assistance for an occupational injury and illness, medical records need to be carefully reviewed. The professional opinion of the physician (or other licensed health care professional) is a key issue not only in recordability, but also in establishing safety culture. If your company has set up care seeking from the lowest bidder, or the clinic that is closest, for only those two reasons, you may be dissatisfied on a variety of fronts; however, both of these items are important factors to consider when choosing health care professionals. We will talk about choosing health care services in more detail later, but let's get back to the records. Once care is given, records should be forthcoming.

The employer should immediately receive a written work status slip that spells out the injured worker's availability for work, work restrictions, medications, and other pertinent information. But there will likely be times when the safety professional finds that even after medical services have been rendered and possibly the worker has been completely discharged from care, a medical record will magically appear through the mail. (Sometimes this record will magically appear with an invoice of some type, but we will save that discussion for later also.) When the medical record arrives, someone needs to place it in the employee's medical file. But before going into the file, someone needs to look at this record and determine what transpired. I have often been surprised at what these records contain.

One item to closely examine is the statement that discusses how the incident occurred. The reviewer will likely find, as I have found, that some workers may give the medical professional different information regarding the accident occurrence. You may learn that your worker was doing something different than what was indicated in the accident report. Whatever discrepancies you find between the medical record and accident report should be resolved and used in the case follow-up.

As mentioned during the earlier example of the metal sliver in the eyelid, you may find that the final report documents that services performed were different from those reported or anticipated. Therefore, an adjustment might need to be made on the OSHA log to remain compliant. The reviewer may find other valuable information such as that the ailment was not work-related. Besides this, the reviewer may find that the worker chose not to show up at his therapy sessions or his final evaluation or chose not to go to a recommended specialist or a variety of other information that the reviewer, as the employer, should be cognizant of.

OTHER TOOLS TO DETERMINE SAFETY CULTURE DEVELOPMENT

Besides using just data that deal primarily with accident occurrences, other tools can give us some indirect indications of culture development. Some of these tools include the following:

- Past OSHA citations
- Safety meetings
- Safety inspections and safety audits
- Job safety analyses
- General assessments

Some of these tools are discussed in detail in the sections that follow. Some of them are discussed in much greater detail in Chapters 10 through 13.

Past OSHA Citations

Past OSHA citations can reveal much about what the safety culture was like in the past. These citations are another valuable tool in the assessment of safety culture. Management must understand that the issuance of a citation is a serious matter. Management is often not willing to spend time and effort exploring, fixing, or defending a citation because the amount of the fine has little impact on the financial success of the company. I am not saying that companies typically do not abate the problem as cited by OSHA. I believe that companies do abate their citations but typically come up with the short-term, temporary fixes rather than looking at the big picture.

Many companies view the whole experience of an OSHA inspection as an almost traumatic experience. I believe that during an OSHA inspection, the management of many companies becomes tunnel visioned toward the goal of having OSHA just leave their premises. If something can be signed or said or done to have OSHA leave more quickly, then someone on the management team will probably suggest that someone sign, say, or do something to make this happen. After OSHA leaves, life at the factory will typically return to "business as usual."

This view is shortsighted. Although OSHA fines and citations sometimes start out as "small potatoes" to many firms, if the situation is left unchecked and the culture continues to lack adequate development, some undesirable results can be forthcoming. Like the iceberg, the OSHA citation may have a hidden section that should be taken into consideration. If your company has received an OSHA citation, this should serve as a warning that your safety culture is in need of development.

Keep in mind that fines can grow by orders of magnitude if the situation gets covered up or does not get adequately resolved. When OSHA returns, even if it is years later, your company takes the chance that the next citation will be much more serious and costly. I will not go into details in this area except to say that *willful*, *repeat*, or *serious* is terminology that should be avoided at all costs. OSHA citations in general should be avoided at all costs. OSHA inspections should be taken with the utmost seriousness.

Besides fines growing as violations continue to be cited, OSHA will also publish those fines on their website. Any company's history of OSHA violations becomes a matter of public record. All one needs to have is a little knowledge of how to get around on the Internet, and this information is readily available. It is also readily available to your competitors. I would not be surprised if your competitors use this information in an attempt to discredit your company. If your competitors are successful in showing that your firm does not perform work safely, then this may provide them with the competitive edge to take work that your company may have been awarded. After all, no one wants an unsafe contractor working for them. The liability just is not worth it. Even without your competitors bringing the issue to your client's attention, OSHA citations are just plain bad publicity. Your clients might reduce the amount of work that you are awarded or choose not to deal with you at all.

Receiving OSHA citations can propel your safety culture development in the wrong direction. Your workers are interested in working in a hazard-free environment. Once a citation is received, team members may have less confidence or trust in management's judgment. This lack of confidence can take years to mend, and in the meantime culture suffers.

Safety Meetings

In a well-developed safety culture, safety meetings play an important role. An effective safety meeting will educate, train, and give workers the opportunity to listen, learn, and participate. If your company does not have routine safety meetings and records of such, this is a huge gap in your safety culture. A technique that has been used with success includes holding a five-minute safety meeting at the beginning of each shift. Then on Friday (or one day per week as logistics allow), a 20-minute expanded safety meeting is held. This technique can be extremely useful when the supervisor finds that five minutes is not enough time to cover a particular subject, but that the subject deems discussion. The supervisor can schedule that expanded safety meeting to cover the more in-depth subject(s). Maybe a five-minute daily safety meeting and a weekly expanded safety meeting is not right for your organization. Whatever fits for the organization should be determined and implemented.

Safety Inspections and Safety Audits

Safety inspections and safety audits are two different things. They are related, but they are intended to spot potential problems at two different levels. Keep in mind that not only the frequency of safety inspections and audits, but also who performs them and their depth can tell a lot about safety culture. We discuss safety inspections and audits in more detail in Chapters 11 and 12.

Job Safety Analyses

A written job safety analysis (JSA) can reveal a lot about a company's safety culture. The safety culture assessment should determine how many JSAs have been developed and for what tasks. There should be a JSA for every work task. The more developed your safety culture is, the more JSAs your company will have. We include some OSHA information about JSAs in Appendix A and discuss JSAs in detail in Chapter 7, "Training."

General Assessments

In 1996 OSHA published a document that is considered by many to be a useful tool. The complete text of this document is included in

Appendix B. This document is entitled, "Tools for a Safety and Health Program Assessment." I will talk about safety programs in some detail later in this book, but for now the reader should remember that OSHA uses the term *safety program* almost synonymously with *safety culture*. Although the terms may be used a bit differently, I believe that the document offers some good general guidance. Within this document, OSHA states that there are three basic methods for assessing safety and health program effectiveness:

1. Checking documentation of activity.
2. Interviewing employees at all levels for knowledge, awareness, and perceptions.
3. Reviewing site conditions and, where hazards are found, finding the weaknesses in management systems that allowed the hazards to occur or to be "uncontrolled."

OSHA suggests using this basic technique to perform the assessment:

- Check documentation.
- Conduct employee interviews.
- Evaluate site conditions and root causes of hazards.

In this document, OSHA suggests that the auditor first review the documentation available relating to each element. The auditor should then walk through the worksite to observe how effectively what is on paper appears to be implemented. While walking around, the auditor should interview employees to verify that what is written and what was observed reflect the state of the safety and health program.

If you are the in-house supervisor or safety person you probably do what OSHA has described on an ongoing basis. In-house people can usually tell pretty quickly if plans are being implemented—and to what extent. If you have an interest in the information offered in this document, the bulk of it has been included in Appendix B.

SUMMARY

Determining accident trends is an important factor when assessing safety culture. We need to determine where we are, so that we can plan our moves forward. In order to help us determine how well developed our safety culture is, there are a variety of tools. The tool referred to by most as the road map is the accident log. This log truly is like a roadmap and should be examined carefully. OSHA logs must be filled out properly using current guidelines and acceptable standard industry practices.

The balanced approach is recommended for recordability. Being too conservative can cause your business to potentially lose awards. Being too liberal in your interpretation may be considered noncompliant. Although recordability is stressed and is certainly important, even more important is determining what has caused the accidents in the past. Besides OSHA logs, there are a lot of other tools in the toolbox to help determine the strength of your safety culture. You need to utilize all of the tools in the toolbox when making your assessment.

4
Program Implementation: How to Start

Before beginning any program implementation, the question of cost will surely arise. Estimates will be forthcoming to include fixed and variable costs. For many companies, the cost of program administration adds little extra cost. Administrative duties for these programs are usually assigned to a person or persons already on-staff. These administrative duties are just "add-ons." The workload may shift a bit, some extra time may be expended, but the true additional cost of program administration is typically minimal. The variable costs of programs, such as time spent and awards, should be easy to quantify.

Although an exact cost of any program may be difficult to determine, the projected cost savings of the program should be the main focus. This amount can be quite large compared to program costs. Once the potential cost savings are widely understood, it should become much easier to begin implementing a program to develop safety culture.

After establishing that the potential cost savings in the improvement of safety culture is huge, and that a slip in accident performance can be devastating, we can begin to map out an improvement plan. The logical question is, "Where do we start?" Start by building the safety program by adding strength in *all* areas. Examine and reexamine how the company operates many times over. A logical place to begin this examination is at the beginning.

Because culture development is ongoing at a currently operating business establishment, the reader may wonder how any program can start at the beginning. Unless the firm is in start-up stage, we are always making changes not at the beginning but in midstream. The challenge will always include changing the culture of those team members who have been thoroughly enculturated the old way; however, consider examining the company's methods for screening and hiring new team members. These potential new hire candidates are at the beginning of a new career. So, for the new hire, we can really begin at the beginning. New hires are discussed first. Then we focus and expand our discussion to employees who are currently employed by the organization.

The new hire starts a new job with a clean slate as compared to the long-term employee. The new hire expects change and is more likely to realize that any routine that has been learned will have to be relearned or at least adjusted. The new hire probably realizes that his or her expectations will undergo tremendous adjustment, especially during the first six months or year. If the new hire starts off "right," the probability that a desirable safety culture is being built increases. On the other hand, if the new hire gets off to a rough start, the road can remain rocky for a long time.

So, when attempting to build or develop a strong safety culture, start with a strong foundation. Ensure that your organization is hiring the best available applicants and is building a strong foundation. This strong foundation will not guarantee a strong safety culture, but it is a good beginning that provides a solid base from which to build (see Figure 4–1).

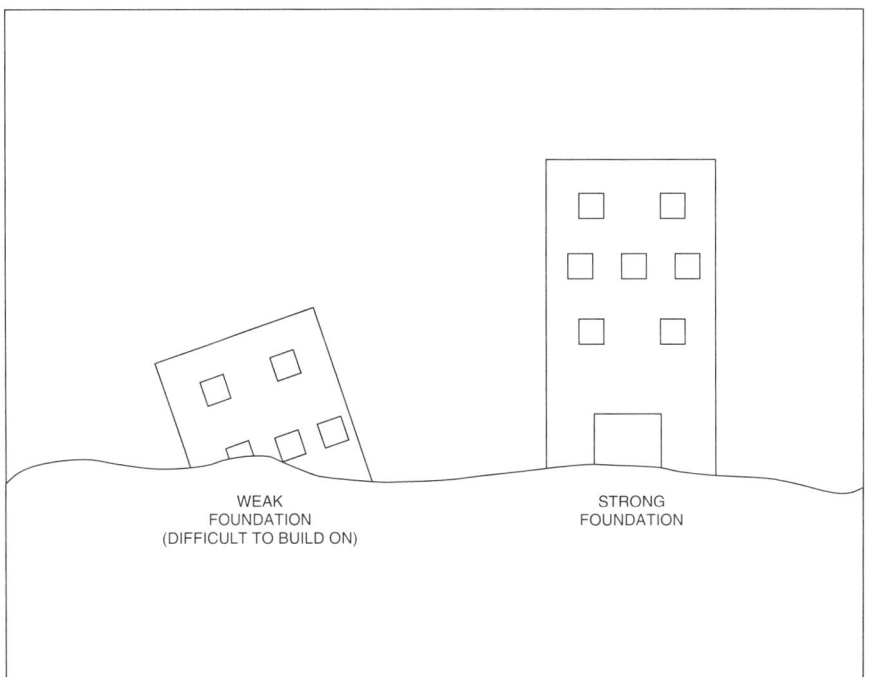

Figure 4–1 A Strong Foundation. Just as within the building industry, a strong foundation provides a solid base from which to build.

The new hire may have a clean slate as far as the new job is concerned, but even if the open job is not the applicant's first "real" job, he or she does not come without some baggage from previous employers, school experience, or general life experiences. For our discussion, the new hire is going to have a cleaner slate and be more receptive to change than the seasoned veterans with high seniority who are set in their ways.

Information will be offered that, although related to safety culture, may not fall under the responsibility of the safety person. These are human resource department issues. If the organization has a human resource department, we suggest that the safety departments and human resource departments work together on screening and hiring. Everyone in your organization should understand the importance of and their roles in hiring new members for your organization. After all, when keeping in mind the Hawthorne Effect, the greatest resource that any organization has is its people.

One must also be extremely vigilant and recognize that various legal requirements must be adhered to during the hiring process. The reader must be advised that before adjusting any internal hiring practices, a qualified human resource professional and legal professional should be consulted to ensure that any changes made fall within acceptable guidelines. Also, some of the information and suggestions offered in this book may not fit every organization. The reader should examine all of the information and then decide which option best suits the particular organization and which information is less likely to be useful.

SCREENING NEW HIRES

Building a safety culture is like constructing a building. Both a strong building and a strong safety culture begin with a strong foundation. Taking adequate time and effort in the hiring process will pay off down the road. Bringing on the best possible people to work for your organization is like forming a strong foundation. The following points are key to an effective hiring process.

THE ROLE OF THE RECEPTIONIST

Train the receptionist (or contact person) to look out for and advise hiring managers regarding any "walk-ins" who indicate they have experi-

ence. The first person that outsiders see when they venture into your company's front door is usually the receptionist. If you have the type of business that hires labor and/or semiskilled persons to perform work for your company, they will probably deal with your receptionist. We recommend that your receptionist be observant if someone turns in an application while the receptionist is on duty. The semiskilled or laborer-type worker may have difficulties filling out an application because of a language barrier. This person may not understand the questions and, therefore, turn in an application that gives less than a favorable impression to any hiring manager.

Your receptionist or contact person can assist by recognizing language barriers and encouraging applicants to take the application with them and get help from someone who understands the questions and can accurately relate them to the applicant. If your organization frequently encounters applicants from a particular nationality or country, make arrangements to have your application printed in as many languages as you feel are necessary.

If the applicant mentions to your contact person that he or she has certain skills or has performed a certain task or tasks that your organization could probably use, the applicant should be encouraged to place that skill or knowledge on the application. If the applicant is unable to write or otherwise include this information, the contact person should write a note to the hiring manager noting the applicant's mention of special skills that have not been recorded on the application.

THE NEED FOR A THOROUGH INTERVIEW

Thoroughly interview every applicant. Every applicant should be carefully considered. Get at least two managers involved in the interviewing process—even with lower-echelon positions. You might try one manager first; then another manager joins in to assist in the process. A bit later the first manager departs, leaving the second manager to interview with the applicant one on one.

Once the interview has ended, the two managers should compare notes. The notes can be in a variety of forms. A comparison of perceived skill levels with task requirements should be considered. During the interview, a determination should be made about how well the applicant would probably be able to perform the job. In order to do this, a well-developed job description should be available for each job. If the job is specialized, the description should be specialized. If the job varies, a description of the job and percentages of each type of task should be part of the job description. Before the interview, the application should be reviewed and potential job opportunities that might match the applicant's abilities should be considered. Each job has certain duties and responsibilities attached to it.

The applicant should be rated (numerically, if possible) by each of the interviewing managers to determine how the applicant might fit in with the organization. There should be some justification of why the interviewee received a particular numerical rating. This numerical rating is similar to being a judge at the Olympics. There are certain "hot buttons" that the hiring manager wants the applicant to push. If the applicant can effectively communicate that he or she has the experience that the hiring manager is looking for, then the numerical rating should reflect this fact. Likewise, if the hot buttons are not pushed, the numerical rating should also reflect this situation.

A minimum of two managers performing interviews is recommended because, in general, two heads are better than one. One person may be able to ask a question or perceive an answer that the other manager does not. When only one manager conducts an interview, the possibility of questions not being asked or forgetting to delve into a particular subject is more likely. Some managers may like to take notes during an interview. During the note-taking process, something may be lost or overlooked because it is difficult to take notes while someone is talking.

If you are an interviewer, make an effort to listen intently during an interview. Ask questions and allow the interviewee to answer them completely. How many interviews have we all been on where we do not get the opportunity to express ourselves fully because the interviewer did all the talking? I have been on numerous interviews where I did not get the chance to say much because the interviewer was so focused on telling me about the job and the company that there wasn't time to share much about me at all.

The interviewer needs to be a good communicator. This communication is supposed to be a two-way street. The interviewee needs to be given an opportunity to get to know the company, but the company really needs to get to know the interviewee.

After the conclusion of the interview, and after the interviewee leaves the premises, the two or more managers who performed the interview should get together and compare notes on the applicant. If they are in basic agreement, good. If they are at opposite ends of the spectrum, they need to determine why these differences occur. The general ratings are usually close but don't mirror each other. If there are extreme differences in their ratings, a redefinition of the job and review by the managers is in order.

MULTILANGUAGE APPLICATIONS

Have multilanguage applications available if appropriate. We briefly mentioned the need for multilanguage applications earlier, but remember that after or during an interview, it might be discovered that your contact person did not realize that an application in a language other than English would have been more appropriate. If this occurs, consider suspending the interview and reconvening when the applicant has been given an opportunity to fill out an application that he or she will be able to fill out most completely and accurately.

It is important to give applicants every opportunity to give accurate and complete information. There are a variety of reasons why this might not happen. For instance, if your applicant is asked whether he or she wants an application in English, that might give the applicant the idea that the company prefers to have only English-speaking, -reading, and -writing persons within their employ. Or possibly that the applicant has a lot of pride and believes that his or her command of the English language is better than other persons might perceive it. So, inadvertently, this applicant will tell you that his or her command of the English language is acceptable, when it might be advantageous to both the applicant and the company if the applicant filled out an application in a language other than English.

Remember that some applicants may have experienced workplaces that may not readily accept non–English-speaking persons or have job openings that the company believes require good communication skills (written and oral) in English. They may have been turned down, possibly on more than one occasion, because of a language barrier or an inability to communicate in one way or another.

An interview is one of the first steps in the hiring process. With good hiring practices, your company can hire the best available people. These people will become part of your firm's developing safety culture, so you should take adequate time and effort to ensure that the best possible applicants are found to fill jobs at any level. The interview can reveal a lot, but it is only a first impression. First impressions during an interview may not indicate a person's potential for long-term employment success.

I believe that long-term success depends more on the person's willingness to learn and determination to succeed.

Let's not focus too much on what we learn or do not learn during an interview. An open mind needs to be kept no matter what the results are from the interview. The interview is just one of the first steps in what is hoped to be a long and successful career. Successful careers can have dubious beginnings. The greatest basketball player in recent history, Michael Jordan, was cut from his sophomore basketball team. His coach then suggested that Jordan take up a different sport because he believed that Jordan just did not have what it takes to make the team. Let's keep this in mind and not prematurely "cut" a worker out of the team because of a less-than-glowing interview.

REFERENCE CHECKS

Check references thoroughly before extending a job offer. As mentioned previously, interviews are a useful tool, but they are not necessarily the best way to determine an applicant's background. Besides the application and interview, one must perform a thorough reference check. References given should be both personal and business. Personal references can be almost anyone, but business references should include co-workers, bosses, and other persons who know the applicant and have worked with him or her on a professional basis.

If your company possesses a human resource department that performs these types of checks for you—great! But even if this occurs, the hiring managers should perform follow-up reference checks themselves. Performing a reference check is a good way for the hiring manager to get to know the applicant. It allows the hiring manager or department manager to have more confidence in the choice of new hire.

Do not be surprised, though, when you run into a situation where the personal or business listed reference either does not recognize the applicant or recognizes the applicant but offers neutral or even negative comments about the applicant. At the same time, the applicant believes that the person will give an outstanding referral. This situation is certainly not the norm. Although it has not happened to me often, it has happened on several occasions. Typically, when you are checking references, you are in the final stages in the hiring process. So, an unfavorable reference can put a cloud over an applicant before an offer is forthcoming.

This may be a surprise to some people, but experience has shown that some applicants provide references that can be less than complimentary. One would believe that any listed reference will give a positive review of the applicant. After all, if you were applying for a job, and your prospective employer asked you to supply names of persons who can "vouch" for your expertise, one would tend to believe that any references that the applicant listed would:

- Be familiar with the applicant.
- Be able to describe the relationship between the applicant and reference.
- Be able to confirm information contained in the application.
- Provide interesting anecdotes or information that can give reviewers insight that they did not get from the application, interview, or otherwise in the hiring process.

I recently received a phone call from someone checking references of an acquaintance of mine. This particular acquaintance had a Ph.D. and was interviewing for a professional position. The person who called me was a retired U.S. Army colonel who would be the applicant's department head. Before receiving this call, I had been advised by my friend (the applicant) that he was looking for a new position and that he would be using me as a reference. So, when the retired colonel contacted me, I was not surprised. I was surprised, though, when the retired colonel asked me to relay to him an endeavor that the applicant and I shared that the applicant would not have shared with him during the interview, but that would give the retired colonel some insight into the applicant's character.

I am not sure if this question is on a list of approved questions to be asked, but it surprised me; I had never had anyone ask me that type of question when checking references. According to the retired colonel, their human resource department typically checks all references, but the colonel always preferred checking the references for anyone who was being considered for hire in his department.

The other end of this spectrum is when you get a phone call from someone checking references that comes as a complete surprise in that the applicant has not mentioned that he or she was looking for a job or never called and asked your permission to be used as a reference, personal or business. It shocks me a bit when I get these calls. I hope that the person who is calling cannot sense my feelings, especially when I must continually ask, "Who was it that used me as a reference?" This situation is less than desirable and can also give the hiring manager some insight regarding the potential new hire.

BACKGROUND CHECKS

Perform background checks before extending a job offer. Background checks can be useful tools during the hiring process. The hiring manager must recall the statement that is seen in many company prospectuses. If you become interested in buying certain financial instruments such as stocks or mutual funds, you might ask for a prospectus before making your purchase. Most of these documents contain disclaimers that have this common message: "Past performance is not necessarily indicative of future results."

Background checks can prove worthwhile in some instances. What first needs to be determined are the job responsibilities for the open position. Determine the responsibilities that this applicant will hold in the near future, and then determine what aspects of the applicant's background should be checked.

If the person will be issued a company credit card, you might consider performing a credit check. If the job requirements call for a person to be bonded or have security clearance, this should be performed. Some companies perform a check to discover pending or settled workers' compensation claims. If the applicant might be expected to drive a vehicle, a Department of Motor Vehicles check on the applicant's driver's license may be in order.

The amount of information available regarding all of us is astounding. We believe that the use of background checks should match the job and, of course, fall within legal parameters. Employers should consult with their legal experts and human resource experts to ensure that any checks on applicants' backgrounds are done in accordance with all legal requirements and are not used to illegally keep from hiring an applicant. Most important, remember that past performance may not be an indicator of future behavior.

WORKING WITH SEASONED VETERANS

New program implementation and the seasoned veteran can sometimes mix as well as oil and water. Just as the new hire is expecting change, so can the seasoned veteran be unwilling to change. The situation can become even more tenuous if your company culture has tried to implement programs in the past that were not successful. The less success that new programs enjoyed in the past, the more likely veteran team members will be to resist acceptance of a new program.

This cycle is difficult to break. The company's current culture has been developing, along with the safety culture, ever since the company was formed. When team members perceive that new programs do not work, the reasons for the past failures must be examined and rectified. I prefer meeting in small groups. Take, for instance, one department: Find a convenient time and place where this group can gather with a member of top management and a small group of others who are instituting the change.

The group should have no more than 25 members. If your departments are larger than 25 members, split up the groups into appropriate sections so that no more than 25 people attend these meetings. Then a well-orchestrated "fireside chat" is in order. Top company officials should not only be present when these meetings take place, but they should also participate during the meetings. Some of the key issues to be discussed are as follows:

- Past practice
- The need for change
- Competitive pressures
- The importance of safety culture to both the company and the team member
- The support that the new program has from top management
- What management expects team members to do (as far as changes) in the future
- A vision of where these changes will take the company in the future
- Other pertinent issues

Once the commitment is in place and the plan for culture development is fully developed, holding a meeting similar to the fireside chat outlined previously is in order. I must stress not to begin the transition until the proper foundation has been laid. If you have this meeting without developing an adequate plan, the likelihood for success diminishes. In later chapters, we will talk about how to get both the new hire and veteran team members on board.

SUMMARY

Establishing the need for a change and getting started with the implementation of that change can be a difficult process; however, once the decision has been made and the task of building a firm's safety culture kicks off, a determination of where and how we should begin must occur. That can be easier said than done because our answer to this question is that we need to begin *everywhere*. The way in which things are done in almost every department should be assessed for safety culture. Top management must assist with the kickoff and stay involved throughout.

Besides some thoughts on how to get started, this chapter emphasizes the importance of hiring and screening new hires. It provides detailed information about what to do during the screening and hiring process so that your employee base consists of the best possible new hires. If you can successfully hire and keep the best qualified candidates for any position, then you will be able to build your safety culture starting from a strong foundation. This chapter also touched on some initial steps to take to get the veteran team member on board and believing in safety culture.

5
Medical Surveillance/ Drug Testing

We expect and require that members of our team be physically able to perform work tasks. In addition, team members must also not be under the influence of drugs or alcohol. Although the physical capability of a worker is typically a management concern, in recent times the mental capacity of a worker, or the potential that a worker is being influenced by drugs, has become a more widespread concern.

The Occupational Safety and Health Administration (OSHA) requires that employers provide a workplace free from recognized hazards. One form of hazard that workers may be exposed to is the unsafe act of a fellow team member. This exposure, which affects all of us, seems to have gotten a lot of publicity lately. Most recently, a tremendous amount of focus and publicity has been given to workplace violence. Violence in the workplace is a tremendously complicated subject on its own. Although we do not discuss workplace violence in this book, business owners/managers need to address this subject. One program that an employer can implement to help alleviate workplace violence is a drug testing program. Because medical surveillance and drug testing can share similar logistics and both typically involve health care professionals, both subjects are discussed in this chapter.

Both medical surveillance and drug testing are areas that your team members will probably be sensitive about. This chapter discusses general principles that have been field tested to be effective. Legal expertise and human resource specialists should be consulted to ensure that any programs your company institutes are administered fairly, comply with current guidelines, and avoid legal pitfalls.

MEDICAL SURVEILLANCE

If we expect our workers to adequately perform their job duties, we need to determine that they are physically able to perform the task expected. This section outlines the following tasks necessary to achieve a successful medical surveillance program:

- Provide a comprehensive preemployment physical.
- Establish and maintain a good working relationship with a local clinic or medical provider.
- Provide annual follow-up screening.
- Make exit physicals a requirement.

Provide a comprehensive preemployment physical. This sounds great, but what is a comprehensive preemployment physical? Unfortunately, this question can be difficult to answer. In many cases, it will take some time and effort to make this determination. Before attempting to determine what makes up your comprehensive physical, we need to determine what that worker will be doing or what tasks the worker will be expected to perform in the near future. The term *comprehensive* varies substantially and depends on job requirements.

If the worker will be driving a truck on the public highway, you will probably have to adhere to U.S. Department of Transportation (DOT) requirements. These requirements include vision, hearing, and a variety of other health monitoring tests. DOT rules also detail drug testing requirements. All of the medical and drug testing requirements need to be adhered to in order for the company to be in compliance with the DOT. If the worker is not expected to drive a truck on a public highway, then the rules for what type of physical or drug test should be used are not monitored by DOT.

If the worker will be expected to wear respiratory protection, your physical exam will probably include a pulmonary function test, chest X-ray, and others. If the job requires that the worker will use self-contained breathing apparatuses (SCBAs) or supplied air, the physical exam might include an electrocardiogram (EKG) along with the other tests. The National Institute of Occupational Safety and Health (NIOSH) has published a variety of information regarding fitness for duty requirements. Attempting to understand these requirements can be difficult for persons who are not familiar with the medical profession. Therefore, most companies work with an occupational physician (Oc Doc) to determine fitness for duty requirements and related issues.

Lessons Learned: Preemployment Physicals

For many years, I have worked with companies that required yearly physicals from all field workers as mandated by the Hazardous Waste Operations and Emergency Response (HAZWOPER) standard. If a worker left the company, he or she was expected to get an "exit" physical. The worker could take the results from this exit physical and go to the new employer, present the exit physical, and use this as a preemployment physical for the new job. Likewise, the companies I have worked for have routinely accepted the exit physical from competitors as preemployment physicals for our new employees. This appeared to be a sound practice because it worked out seemingly well for many years.

Also, most firms have a philosophy that no one will begin work until having successfully passed a drug screen and physical. As stated previously, the legal aspects of the hiring practices should be approved by your company's legal and human resource department. This discussion is not an attempt to advise a firm to adopt any hiring practices that are not within legal guidelines.

As with most companies that are very busy, when they find a worker they wish to hire, the hiring manager wants to do that hiring *now*—not later, and not waiting three days to two weeks to get laboratory results back. So, even though there was a policy in place not to hire anyone until successful completion of a drug test and physical, if we were busy, some corners would probably be cut. If the worker that we were interested in had a fitness-for-duty slip that was less than a year old, and had passed the drug screening, it was standard shortcut practice to put that person to work. The new hire would go to work before blood work and physical results from our company's physician were completed.

In this situation, this particular team member had less than a week on the job when the complete medical results were received. The results indicated that the worker was not fit for duty. What seemed to make the situation even more delicate was that this worker had impressed his supervisors with the work he produced. He fit in well and seemed to be able to perform the work handily. It was a sad moment when his supervisor informed him that he was not fit for duty and why.

After a little bit of research, it was discovered that there was a tremendous difference in the battery of tests that the physicians

(continued)

> administered to show fitness for duty. It was like comparing apples to oranges, yet both physicians would provide a fitness-for-duty slip that was worded almost identical to each other. So, get comprehensive preemployment physicals for all workers. Do not allow a worker to go to work until *your company physician* agrees that this worker is truly fit for duty.

Establish and maintain a good working relationship with a local clinic or medical provider. A good relationship with the medical providers should be ongoing. Questions pertinent to fitness for duty should be left to the discretion of your company's physician. You may find that philosophies of physicians vary widely. Because of these varying philosophies, it is important that you spend the time and effort to choose a physician whom you think will provide your company the type and level of care to workers that correlates with the company philosophy. Later we discuss return-to-work policies and working within restrictions. The relationship with the medical provider can make a big difference in the effectiveness of your medical program.

Different locales can have large differences in the way occupational (workers' compensation) cases are handled. No matter what the requirements are for your state, the quality of the entrance physical should be high. The relationship that the physician established with the worker on a preplacement physical helps set the tone for the safety program and building an effective safety culture in your organization.

It can be difficult to establish a good working relationship with a medical provider, especially in a rural environment. In rural America, there are often no or very few occupational physicians within a reasonable travel time or distance. Even in large cities, finding physicians who reflect the values of your organization can be challenging. You may not be able to locate a physician specializing in occupational medicine, or an Oc Doc. You may not be able to find the medical providers who match the values of your organization, and will have to resort to the best available. No matter whom you choose as a medical provider, try to make sure that the provider has a reputation for being thorough and knowledgeable.

Work with the provider to help give good service to employees. It is not advantageous to send someone for a preplacement physical and then have them spend a half-day or more getting a physical examination completed. If there are busy times or slow times at the physician's office, find that out. Work with the provider to arrange for appointments (at offpeak times) for preplacement physicals. The type of service and the time it takes to get the physicals done can leave the worker with a lasting impression about the medical facility that we hope is positive.

If the worker is injured at some point in the future, that worker will have confidence that he or she is getting appropriate treatment for the injury. If the worker believes that the medical facility is full of noncaring, incompetent, inefficient persons, it will add to the likelihood that the worker will not trust anyone in that facility. This lack of trust is to be avoided at all times and could become an important issue if the time comes when the team member receives treatment from that facility in the future in connection with an on-the-job injury.

The chosen medical facility should be evaluated at regular intervals. This evaluation should be in the form of a field trip to include a face-to-face meeting. The safety professional or other management representative should get to know the medical professionals and work closely with them on medically related issues. Any questions that arise should be adequately answered. The program should be reviewed annually or at least regularly. Any medical forms that are in use should be reviewed to make sure the latest and best versions of the forms are being used. The battery of testing that makes up the physical exams should also be reviewed. The types of questions that should be considered include the following:

- Is the program up-to-date when compared to modern medical practice?
- Should further testing be added to bring the program up to par?
- Should testing be eliminated that may no longer be needed or considered overly conservative? (e.g., yearly X-rays for young workers)
- Is there a better way to get the medical treatment accomplished that will be good for the worker, the company, and the medical provider?

Note: Further discussion about medical providers is offered later in this book, particularly when we deal with light-duty programs and working with restrictions in Chapter 14, the supervisor training chapter.

Provide annual follow-up screening. For companies that believe their worker exposure to chemicals is negligible and where respirators and personal protective equipment (PPE) are not used, and that the work they expect workers to perform will not be physically taxing, preemployment or annual physicals may not be required. For the many companies that deal with chemicals or have their workers use respirators, or whose work includes strenuous physical labor, annual physicals have been a way of life.

Whether a physical is a company requirement or not, workers are typically encouraged to supplement the company-provided physical with their own personal physician's evaluation. Even though personal physicals are encouraged, many workers substitute the company-provided physical for a personal physical. There may be some logic in this philosophy after all, if the medical provider is performing a comprehensive physical, because the information obtained is helpful both on and off the job. One should consider not only seeing the company physician on a yearly or as-required basis but also visiting a personal physician regularly. You should be able to obtain the results from your company physical and provide these to your personal physician. The personal physician will probably review the results obtained from the company physical and concentrate on any areas of potential concern.

Lessons Learned: Medical Information Sharing

> I have worked with various members of management on HAZWOPER jobs where workers obtained physicals from the company-approved physician and their supervisor would not share the results with the workers. This situation went on even though the HAZWOPER Standard specifically states that the information will be provided to workers who request it. Unfortunately, one of the local front-line supervisors insisted that because the company paid for the physicals, the results were the company's property, and he refused to allow the information to be shared with the workers.
>
> In my opinion, not sharing information such as physical results with the team members who took the physicals is a poor idea. This supervisor may have exhibited poor judgment when not allowing medical results to be shared with team members. Information sharing and developing a feeling of trust between team members and management is an important part of safety culture development. Even if sharing the results of physicals was not specified within the HAZWOPER Standard, sharing personal medical results of any kind with the team members is recommended. In general, just sharing information with the worker will enhance culture development.

Are annual physicals a bit too extreme for your organization? Maybe the expense is too great a burden; maybe only the field people or workers should go annually, and the supervisor and upper-echelon people should go every two or three years (or even less often). The frequency of physicals when not mandated by standards is something that demands careful consideration. When companies are in austerity modes, things such as physicals, incentive programs, and safety programs get the same hard look as other areas. Even when a company is not cutting costs, maybe annual physicals should not be required for all workers.

Perhaps varying the frequency of certain sections of the physical makes sense for your organization. X-rays were mentioned previously. Does your medical provider feel that X-rays should be given to all workers of all ages annually? There are a variety of views on this subject. Some physicians believe that the amount of radiation workers are exposed to should be limited as much as possible, especially for workers under 40. The Nuclear Regulatory Commission has different amounts of allowable radiation depending on age.

If you do decide to vary different sections depending on physician recommendations, this may have an adverse affect. Complaints received under similar circumstances include workers believing they are not being given adequate care, to older workers stating that the X-rays will be used to force them into retirement early, or are another form of age discrimination. So, whatever the reasons are for adjusting the contents or frequencies of physical examinations, we recommend that you communicate this rationale to your workers and to the medical provider. Encourage workers to discuss their concerns with their supervisor, human resources personnel, the medical provider, their own physician, and whomever they feel comfortable talking with. Be honest and open with them, and address all of their concerns.

Make exit physicals a requirement. Workers who are interested in changing employers or quitting seem to do so with more courtesy than in the past. Thoughts as to why this is happening include that prospective employers are getting better at screening prospective workers. It seems more likely now that an employer would check references and the circumstances of why workers would leave a job than occurred in the past. In the 1980s, there was a popular song entitled "Take This Job and Shove It." For those of you who may not be familiar with this song, it is a story about a man who is fed up with his job and "tells off" his boss as he leaves the business establishment. My recollection of the exact words of this song include: "Take this job and shove it. I ain't working here no more."

Obviously, this worker (I believe the artist's name is Johnny Paycheck) does not give notice, nor does he leave on the best of terms. It is doubtful that his supervisor would give Johnny Paycheck a positive reference. It does not take long for workers to realize that leaving a job under strained circumstances can lead to a rougher road ahead when looking for a new job.

So, for modern workers, the likelihood that arrangements can be made to obtain an exit physical has improved. From the beginning, all new hires should be advised that it is a company requirement that any person who leaves the company (for any reason) must undergo an exit physical without exception. This exit physical will provide a medical record of the person as he or she leaves the job. The exit physical should be as comprehensive as the preemployment physical. If the guidance offered earlier is followed, completing an exit physical should not prove to be an insurmountable difficulty. In fact, many employees might be able to save their prospective employer a little money by using the exit physical that their present employer pays for as a preemployment physical for their new position.

Note: As mentioned earlier, using another company's exit physical as a preemployment physical is *not* recommended; however, many employers still follow this practice.

Dealing with Disgruntled Employees

You may come upon a worker who is simply disgruntled. This worker will find fault with anything that has to do with the employer. His or her list of complaints will be extensive, and the last thing that he or she will do willingly is to get a physical from a company-sponsored physician. Dealing with a disgruntled individual can be challenging. It is hoped that the disgruntled worker will be a rare occurrence within your organization, and you will never have to deal with it. But if you do, keep these points in mind:

- If you are approached, and you are alone, try to keep conversations calm.
- Immediately get on the telephone and request that a co-worker come to your office (as backup) and be part of the event.
- If possible, don't even *begin* a confrontation until you have another member of management with you.
- Point out to the disgruntled person that it appears that an impasse has been reached and that the conversation should be ended because it is unlikely that further discussion will reach a solution at this time.

Workplace violence has been in the news often recently. Although this subject is important, this book will not address workplace violence in more than a general way; however, numerous sources discuss this subject in great detail. If it is taking place in the workplace and has a potential to harm workers, OSHA is typically interested. My hope is that with a well-developed safety culture and keeping in mind the Hawthorne Effect, workplace violence issues should be eliminated or at least minimized.

We have gotten a bit off track with our discussions of workplace violence. Let's get on track and summarize exit physicals. I believe that one should offer all persons leaving your employ an exit physical. Let them know during their initial orientation that an exit physical is company policy. Ensure that the offer for an exit physical is consistent for all workers. If the person refuses to take the exit physical, make sure that you document the fact. Include with your documentation any reasons mentioned for not taking the exit physical.

DRUG TESTING

We expect our team members to be physically and mentally capable to perform their assigned work tasks. We talked about physical capabilities earlier in the chapter; we now discuss mental capabilities. Many years ago, I was interviewing for a job with a large chemical distributor. I was chosen for the job and accepted, but I was not asked to take a physical or a drug test; however, before I could start, I was required to take a battery of tests to determine my personality profile. These tests were supposed to give management insight into my ability to perform my job in the long run. Basically, this company was trying to determine if I was mentally able to perform expected work.

Employers would like to ensure that the team members they are hiring will "fit in" and last for the long run. Although I cannot agree that personality testing such as I described is truly a useful tool to make these determinations, I do believe that drug testing has its place. I believe that drug testing should be used to the extent legally acceptable and warranted by the work activity. Drug screening or testing goes hand in hand with the company's medical surveillance program. Having an effective drug screening program is like buying culture development insurance. Drugs and alcohol continue to contribute to accident occurrences. If we expect to build an effective safety culture, we will be building this culture with workers who are screened "drug free."

Depending on the type of business that your firm is in, and the demographics of interviewees, it may be advantageous to perform drug screening before sending new hires for a physical. For instance, if a firm hires general labor-type positions and already has a prescreening drug policy in place, the company has probably spent (or wasted) a large amount of money on preemployment physicals only to realize that the worker will not be offered the job. On the other hand, operations people, when they are hiring, are in a hurry to get their choices hired and functional. After all, we need that person to do a particular job, or we would not be hiring them. Sending a person for a drug test separate from a physical may appear to be an inconvenience, but it can turn out to be a necessary and useful step.

What happens to the team member who tests positive during a random or annual drug screening? Or better yet, what happens to a potential new hire who tests positive for illegal drugs before starting work? Both of these questions are excellent, but neither one is discussed in this book. As with many other related subjects, volumes of material are available that discuss the answers to these questions. These situations can be extremely sensitive and complicated. Significant human resource and legal consultation needs to take place before answering these questions.

Lessons Learned: Surprising Drug Screen Results

> As a safety person who has held positions at various management levels, I have been involved in the hiring practices and drug screen/physical practices for some time. I have worked with companies that asked their current employees to reference new hires to ensure that they were hiring people who were drug free and good workers.
>
> Management for this firm felt that a reference from a current team member in good standing far outweighed running ads for employment or other alternatives. The reference method ensured that all prospective new hires are told that they would be asked to undergo screening for drugs before being hired. If the prospective team member was using drugs, one would think that after being informed that drug testing would be required, the prospective team member would either find a reason not to pursue the job or avoid the use of drugs for a sufficient amount of time so that he or she could at least past the initial drug screen and get the job.
>
> Believe it or not, many people screen positive for drugs even after being informed that a preplacement drug test is required. Some of the people who tested positive for drugs have surprised me. Some of these people also surprised the people who recommended them for the job. Being a student of human nature, I thought that I would have a good feel for determining whether the drug screen might come back positive on a certain candidate. I have been surprised quite often.

On a construction site in the Midwest, the company I was working for had to hire a variety of construction laborer-type team members. The job was a construction-type project that would last approximately one year. The employment opportunity could last for the duration of the project. The hourly wage being offered was considered an excellent level given the state of the economy in this general area. There was no problem in finding people interested in applying for the job openings. The problem came in finding people who could pass the drug screen.

I was personally surprised at which prospective team members would pass the drug screen and which ones would not. Before new hires would be sent for a drug screen, at least two management types would examine their application and conduct an interview. It seemed that some of the most unlikely candidates would pass the drug screen and some of the ones who were thought would have nothing to do with drugs (and typically told you so) would test positive. Various members of the management group were also surprised at who tests positive with preemployment drug screening. As you can imagine, it can prove embarrassing when someone you refer for a job tests positive for drugs.

DRUG SCREENING AND TIMING

One might ask, "What's the big deal? How long can it take to perform a drug test?" The answer to this question ranges from near instantaneous to a week and a half or more. Almost instantaneous results can be obtained in a couple of ways. One more common way to get these almost instantaneous results comes from performing the tests in-house with the latest "test kits" offered by various manufacturers.

These high-tech testing devices may present a few more difficulties than one might realize. First, someone needs to have the training to properly administer an in-house program using the "test kit." You must determine who in your organization would be this qualified person. If an in-house nursing or medical staff is available, this would be a likely place to find this qualified person. If you do not have an in-house medical or nursing staff, you might consider using your human resources manager. If you do not have a human resource person, maybe it's the safety person.

Whomever this person turns out to be, this task must be performed meticulously. Besides the precision needed when performing this test, any and all information regarding the outcome of any such tests must be kept confidential. During the test itself, a possibility exists that a bodily fluid may make contact with the tester.

Because of the nature of this type of testing, it is imperative that the person doing the testing have extensive training in a variety of different areas, including bloodborne pathogens. Although some companies have an excellent in-house drug testing program, I prefer to have this service performed by an outside source rather than in-house. As we continue, I will give you some reasoning for my beliefs. Let's first look at the failure of a drug test.

When someone fails a drug test, whether this is a preemployment test or otherwise, the failure of a drug test can be a sensitive situation. The person who has tested positive will often go into denial. Sometimes the denial is vehement. This denial situation can be unnerving, especially the first couple of times one experiences it. If you are performing the screening through an in-house person, this person can be the center of some harsh attention.

When an applicant fails a drug test that was provided in-house by the company, those person(s) who performed the testing need to have thick skin. They will probably bear the brunt of some harsh criticism when facing those who tested positive. This situation can be avoided or minimized when an applicant who does not pass a drug screen is informed through an outside source. This outside source can provide a useful buffer. Using an outside source can also be a buffer between whoever recommended the applicant and the hiring company. This can be an important point to consider when one of your current workers made a recommendation and this person fails a drug test.

Another consideration for doing screening in-house is determining where the screening should be performed. Clinics are typically able to provide appropriate facilities, privacy, confidentiality, and hygiene that need to be part of in-house testing. Many companies are not able to provide this type of test. Companies will probably have some type of reception area where people can be greeted and maybe fill out an application. But they may not be set up with a private bathroom with an adjoining hygienic private work area/countertop where the sample can be readied for shipment and the chain-of-custody and other paperwork can be filled out in private.

On the other hand, depending on logistics, your company may already be set up to perform in-house testing, or minor modifications can be made so that in-house testing could easily be made. Typically, the people who are marketing the test kit products that are becoming popular will help you with setting up your facility so that your program gets off to a good start.

Besides test kits, there is another way to get back almost instantaneous drug screening results—an in-house lab. If your organization has an in-house medical staff, having the equipment to perform drug screening in-house may be something to consider. Although these situations are not common, I have been involved with drug screening where the sample was analyzed and results given in less than an hour. The industrial facilities where I have seen this in-house screening performed had a full-time medical staff and employed thousands of persons at a single facility. When you are testing large numbers of drug samples over a long period, running the numbers will determine if this type of situation is a cost savings to your organization.

For most organizations, the two big advantages to performing drug testing in-house are time and convenience. These factors can become large depending on how long it takes you to get back your drug test results using your current method (if you have one) and how inconvenient your medical facility is when compared to your plant (or business) location(s).

For a company to start with no drug testing program and move toward having a program of any type is a large step. Companies that do not have any type of program have probably considered the added value and decided that a drug testing program is not a good choice for their firm. There may be a lot of reasons for not having a drug program, but one should consider all of the options before making a final determination. In most instances, when the work being performed is labor intensive, I continue to believe that some form of drug testing is good for the safety culture.

Within the main drug screening program, there are typically three subprograms:

- Preemployment testing
- Random testing
- "For-cause" testing

The "for-cause" program may be linked to behavior or accident or near-miss incidents. Although preemployment drug testing was discussed earlier, we failed to discuss yearly or planned testing. Companies will require everyone to be screened for drugs when they take their (annual) physicals. Similar to preemployment drug testing, the workers realize that the drug screens will take place in conjunction with physicals. The workers realize (or should realize) when they are "due" for a physical and likewise an announced drug test. Even though the workers realize that their physical will be coming up, people still "fail" their pre-announced drug screening. We will not discuss why workers might fail a drug screen, but instead reinforce the importance of building a safety culture with a drug-free workforce.

If your company decides to implement drug testing consisting of the three features mentioned, a random program can prove to be a challenge, especially for smaller companies, primarily because of statistics. The smaller the number of persons who participate in the random program, the more likely it is that the same person will be chosen twice or even three times before everyone has participated. The person who has been chosen two or three times may feel that he or she has been "singled out" or "picked on." Or other workers may feel that this person has been singled out or that the randomness of the program is "fixed."

A method used to choose participants in the random program that has been field tested to be effective is to assign all of your participants a number. Obtain Ping-Pong balls with numbers (like those used in bingo games) to match participants. Pick the appropriate number of Ping-Pong balls in front of others and have different persons make the pick every month.

Another method to use when attempting to ensure randomness is outsourcing. Your medical provider can probably make some recommendations about who to contact that will administer your program. The fee charged to administer the program is typically nominal.

The U.S. DOT has a variety of information on drug testing for commercial truck drivers. If your company is looking for a proven program, this program has passed the test of time. Although your workers may not be truck drivers, the program has a variety of features that your company might find useful. No matter what method your company decides on, you will probably hear comments if someone is picked more than statistically equal.

The most sensitive area of a drug screening program is the for-cause section. In other words, the workers have acted in a certain manner or have exhibited some behavior to cause someone to believe that they should be screened for drugs. This might include workers acting as if they are intoxicated or under the influence of an illegal drug. It can be difficult for supervisors or other management members to require that workers take a drug test because they observed or think they observed that the workers were acting as if they were intoxicated or under the influence of illegal drugs.

This area can be extremely sensitive. I was recently involved with a case where a supervisor felt that a team member was "high" because he noticed that the team member was slurring his words. It turns out that the team member had a variety of medical problems that were not related to drug or alcohol use. This team member had experienced a stroke several months before, and with a variety of other medical problems might slur words when his prescription medication or therapy sessions were being administered. Once this situation was resolved, it became obvious that the supervisor–team member relationship needed to be bolstered. The relationship between the team member and the supervisor is a crucial factor in the development of safety culture. Any members making the judgment regarding for-cause testing should undergo extensive training.

The U.S. DOT program mentioned previously contains useful information that may help your company develop an effective for-cause drug testing program. Under the U.S. DOT program, the for-cause issue is spelled out clearly. It includes strict parameters for mandatory testing when monitory limits are reached for both property damage and personal injury incidents. It discusses property damage for both your company's property and other person's property. It also gives other guidance regarding personal injury cases.

This program has withstood the test of time for commercial truck drivers all over the country, including legal challenges. Although the principles within the U.S. DOT program are sound, the circumstances at your company may not exactly match those offered in this program. After all, the U.S. DOT program is set up for a workforce made up of commercial truck drivers. If this is not the workforce for which the drug program is being considered, adjustments will be in order.

A similar basis or logic that is used in the U.S. DOT program could be a starting point for other programs. Success and ease of administration can be achieved by defining parameters that constitute cause, and then by making painstaking efforts to train managers who are expected to work under the program, along with closely following up the administration of the program in general. Some programs have been set up to include mandatory drug testing when a certain dollar value of damage is reached. Other programs include mandatory testing for certain types of medical treatment.

For instance, some companies institute a for-cause drug testing program that goes into effect if a worker has an on-the-job injury that requires medical attention. So, if a team member would go to the clinic, this would automatically trigger a drug screen. The medical provider who was chosen is already on board with this company for physicals and preemployment drug screening. In this case, it takes little effort to get the medical providers to administer this service.

This situation may seem straightforward. When someone gets hurt on the job and receives any kind of medical attention, a mandatory drug screen is performed; however, in this situation—as in most cases—things are just not as simple or straightforward as one might think. It is common to run into numerous bumps in the road when trying to administer an automatic drug screen following a job injury that requires medical attention. For instance, how do you handle someone who reports an injury that occurred days or even weeks ago? And what about the worker who is taking numerous amounts of prescription medicine? Or the drug screen that comes back with an indication that an agent used to mask a drug's identity has been identified in the sample?

The list can go on and on, but if you advise your workers when you hire them that it is policy to test anyone who has a work-related injury that requires medical care, and you administer your policy fairly, the chances for success increase substantially. If you do a poor job in any area of your drug program, your chances for developing your safety culture diminish.

SUMMARY

Medical surveillance and drug screening are two programs that every company should seriously consider adopting. One of our company core values should be that team members will be physically able to perform their work tasks and be free from the influence of alcohol and drugs. The basic principle is that workers are supposed to be physically and mentally capable of performing their expected work tasks.

Maybe not all workers will be required to participate in a medical surveillance program. Many companies only require medical surveillance and drug testing programs for those workers who are known to be performing physically challenging work activities or required by government regulation to undergo medical surveillance. Too often, companies learn too late that a team member was not physically able to perform their assigned work task. These situations can be avoided with use of an effective medical surveillance program.

An effective drug testing program can be difficult to administer. This area continues to provide challenges for those administering these types of programs. For many situations, the use of an outside vendor to provide drug testing services is a logical choice. These outside vendors will typically offer full-time workers that specialize in this area and can provide a buffer between the employer and the person(s) being tested.

6
New Hire Orientation

No matter what type of job the new hire is tasked to perform, the new hire should receive *some* type of orientation. Some type of orientation is best management practice even if your new hire is a temporary worker who is expected to do routine clerical work and/or answer the phone.

The type and length of the orientation that a new hire should receive should be performance oriented. That is, the orientation that the new hire receives should depend, to a certain extent, on the type of job that the new hire will be expected to perform. Everyone should be oriented regarding the workplace logistics or layout. Depending on the circumstances, the orientation could include the proverbial nickel, dime, quarter, or more tour. These circumstances usually consist of the work tasks that the new hire will be expected to perform.

Consider Maslow's hierarchy of needs. The orientation must include information that includes taking care of the worker's basic needs. Someone needs to ensure that the new hire is shown the location of the men's or ladies' room, the lunchroom, the vending area, the parking area, and other areas of common interest. Emergency procedures and details about how to use the phone and summon emergency responders should also be provided. The new hire should be introduced to co-workers as the tour is taken.

As mentioned previously, no matter what type of job a new hire team member is starting, some type of orientation should take place. This orientation is an important part of the new hire's career. It gives the new hire the first impression of the company and fellow team members, and it gives management and fellow team members that important first impression of the new hire. We might think that orientation is only for lower-level labor or office help. This should not be the case. Everyone, including managers, should go through an orientation process.

Orientation should stress safety and the team. Whoever is assigned to conduct orientation for new workers should walk around with the new hire and introduce him or her to the other team members. If the workforce is large, it might be advantageous to start in the home department and concentrate on making introductions there. If the company is small, possibly everyone whose path is crossed should be introduced to the new hire. Teaming must be emphasized, along with safety. After all, the Hawthorne Effect should be considered for the new hire, as with all team members. The following two key points should be kept in mind:

- All jobs (and all workers) are important.
- If the job were not necessary, the position would not exist.

These points are valid whether we are dealing with the janitor or the CEO.

OFFICE ORIENTATION

As mentioned earlier, orientation should have many levels or steps. Let's consider the orientation of a general laborer in a company with fewer than 100 people. (We can talk about orientation at a larger company later.) In this situation, when a laborer is hired, the orientation will likely involve an office orientation and a field orientation. The office orientation will likely include a variety of different people from different departments. These people might be a benefits person, the office manager, an administrative assistant, a human resource person, and a supervisor type. For the extremely small company, the benefits person, office manager, administrative assistant, and human resource person might be one person wearing all these hats.

New hires should get a tour of the office facility. The tour should be guided and extensive. For some laborers, this may be one of the few times they see the inside of the office, so the orientation must stress teaming and be positive. Safety must be stressed as a core value to the company. During the office tour, the types of things that an orientation should include might be the location of the men's or ladies' room, the lunchroom, the vending area, and the parking area. You can imagine that the new hire will be quite on edge during the first few weeks of a new career, and especially on the first day. Just finding the office location on time may have been a struggle, and the fact that the new hire has a human to ask questions of can start the working relationship off on the right foot.

As previously mentioned, but important enough to be stressed, the new hire should be introduced to co-workers as the tour is taken. It is unlikely that the new hire will remember all of the new names, but he or she might remember that they are part of an important team and that safety is important to their new company and job.

USING VIDEOS FOR ORIENTATION

The use of videotapes during employee orientation is widespread. The use of video has some advantages over face-to-face, handouts, or other orientation methods. If the video is well done, it will have a script that has been deemed accurate and complete by the reviewers. Giving accurate and complete information to team members pays off now and in the future.

With the advantages that come with the use of videos, many companies tend to "overuse" the tool. In many locations, for many different reasons, videos have become overdone. Some companies have many different videos to cover as many in-plant scenarios; however, one is less likely to find a qualified, live, knowledgeable person who is available to answer questions or even ensure that the video was watched. In this day and age, finding a live human can be a challenge. It is becoming even more popular for phones to be answered electronically. There is less and less chance that you can find humans who answer a phone even after one has gone through the phone maze options. And the lack of a qualified, knowledgeable, live human makes the video presentation an appealing choice for many business concerns.

Videos can use professional actors. These actors can have an advantage over the people who actually do the work in that actors are used to performing on film. They can exude confidence and really sell a concept. Not using professional actors carries with it some disadvantages, one of which is that a person who is not used to performing on film sometimes looks like a novice. This can give the orientation a less than professional look.

Videos can also be produced professionally. Besides having professional actors, they can have professional writers, directors, mixers, musicians, and so on. This can mean that the videotape used for orientation is one that is a pleasure to watch, along with containing accurate and useful information.

The information presented in the video (or the script) can be reviewed to ensure that the content is accurate and complete. The viewers are ensured that they will get accurate and complete information as long as they pay attention to the video. This can be an advantage to the orientation provided by the human, who can develop a tendency to skip over information. This tendency for the human presenter to skip over information increases as the number of orientations that the presenter makes increases.

Videos can be produced in a variety of languages, or produced in English and then have the voice track dubbed in however many languages are necessary to get the information across accurately and completely. The philosophy that everyone must have English as their primary language just does not reflect working reality, especially when dealing with labor-type jobs. So, be prepared to have your orientation videos translated to as many different languages as the persons attending the orientation are likely to speak.

A problem situation arises when the person or persons who are supposed to present the orientation do not show up due to sickness, traffic, tardiness, forgetfulness, or other reasons. Or the person(s) scheduled to perform the orientation shows up but can't do it because they have been stricken by laryngitis, lost their notes, it isn't their turn, they have another commitment, or a variety of other reasons. If you have your orientation on a videotape, it is likely it will be done as long as you can find a functional VCR/TV and a working electrical outlet.

In the previous few paragraphs, a lot of advantages regarding the use of videotape presentations for orientation of new hires have been mentioned. Even though there are many advantages to this method, one needs to ensure that using video as an orientation tool is not overdone. Videos have their place, and I believe in their effectiveness when used properly, but they should not be overused.

Try to limit the use of videos during orientation to 20 percent of the total orientation time. Try to limit the time spent watching videos to 20 continuous minutes. An effective use of the video to consider is to have the presenter stop the video at key spots to emphasize the important points or discuss field changes or other pertinent information not adequately shown in the video. One might consider having a written test administered after the video to ensure that viewers retained important points. This written test should be corrected before allowing the new hire to continue the orientation process.

Consider the new employee (general labor or operations) on the first day on the job. This new hire will likely view a video that provides general information about the company. This general information might include the products and services that the company provides. Or maybe the general orientation will contain information about the history of the company and its key people.

In general, presenting this type of information during an orientation would be considered a good idea. This type of video is as much public relations or sales as it is safety orientation. Having an idea of what the company is all about is a good way to build confidence in a new hire and start developing a team attitude and a safety culture. Remember that the length of this video should be kept to no longer than 20 minutes. If it goes much longer, people have a tendency to "drift." Shorter is better. Consider a 10-minute general orientation as the optimum length.

After viewing the short general video, the new hire might view a more detailed and lengthy department-specific video. This video would show workers performing the same tasks as the new hire will be expected to perform. The portrayal of the work should be authentic. If the work is dirty, cold, difficult, and dangerous, the video should reflect these real-life conditions. If the work is not so dirty or dangerous, then the video should reflect those real-life conditions.

Even though using professional actors on orientation videos has some advantages, I believe that when weighing both scenarios, the culture is better off by using team members for actors versus professional actors. Before choosing professional actors, one should seriously consider using team members or "real" people. The typical worker should be used on the video. It is not a concern that the typical team member might not be in robust physical shape. And do not worry if the typical worker has some holes in his or her clothing or that the work uniform is a bit dirty. Concentrate on the work task and not on the dental work or acting ability of the typical worker.

These "homemade" videos can be an excellent orientation and training tool. I find they work best when a script is not strictly followed. Instead of following a strict written script, consider following these loose guidelines:

- Inform the typical worker that you are putting together a video that will be shown to new hires so that they get a preview of the type of tasks they will be asked to perform when they actually start working.
- Advise the typical worker that you are interested in demonstrating to the new worker the "right" way to perform the task at hand.
- Let the typical worker know that you will be asking questions about the task that is being performed and allowing the worker to answer the questions in his own words without a script.

- Take a couple of practice runs if you need to and play them back to the worker. You may notice that the worker needs to speak a bit louder to be heard or that other adjustments need to be made.
- Make those adjustments and finalize the video.

The typical audio on this type of video would go something like this:

"I'm *Jack* of the company's safety department, and I am going to interviewing *Jill*. *Jill* is an experienced widget operator who has been in this department for __ years.
Jack: Good morning, *Jill*."
Jill: "Hi, *Jack*. How are you?"
Jack: "Fine thanks, and you?"
Jill: "Okay."
Jack: "*Jill*, we came down to see how this department makes widgets. Can you give us a bird's-eye view and some pointers?"
Jill: "Sure, *Jack*. This department takes widget manufacturing from step five through step eight. The widgets are transported to this department from the brake press department. Notice the color and shape of the widget at this point . . ."
Jack: "*Jill*, have you ever had widgets show up here that weren't quite right; I mean, the color and shape were off?"
Jill: "Oh yeah, a couple of times. You have to be very careful and inspect each load to ensure ___, and when this happens you need to contact ___ at this extension right away."
Jack: "Great. Well, how about step six; how does that work?"
Jill: "Well, *Jack*, step six works like this: You retrieve the widget off the line after completing step five and place it into this fixture and then . . . Here are some examples of the ones I did this morning."
Jack: "*Jill*, what if I place them in the fixture like this instead of like that?"
Jill: "*Jack*, that's very bad. If you do that, the widgets won't fit into their shipping boxes at the end of the line."

This type of scenario goes on until you complete your orientation. Don't use a script, but instead get the typical worker to explain the work task in his or her own words. Lead the conversation with pointed questions, but don't put words in the actor's mouth. Take the video showing the ins and outs of the job.

The desired result is that the inexperienced worker or new hire will benefit from the knowledge shared by the experienced worker. When mistakes are made, you want these mistakes to be a learning experience. It is hoped that the mistakes shared on the video by the experienced worker will not be repeated by the inexperienced worker. When we learn and benefit from our mistakes, we will be continuously improving our performance.

This discussion began over the pros and cons of using professional actors/actresses instead of real workers. We then talked a bit about the advantages of using team members rather than a professional actor(s) for orientation. Let's discuss this subject a bit further. First, let's consider the videos that are professionally produced and use professional actors. The views of the work in progress are usually very good, but the acting usually leaves much to be desired. Although the scripts may be accurate and complete, one can usually tell that the actors really have no in-depth knowledge of what they are talking about—it's an act and people see through it.

Something else happens when we use professional actors in orientation or training videos rather than the people who normally perform these jobs. That is, using the professional actors sends a message to our workers that, for some reason, management has chosen the team member to do the work but not to take part in the video. I believe that this decision sends the wrong message and should be reconsidered. Even though management's intention might have been to put together a world-class orientation video, not using their own team members might be giving mixed signals.

I believe that the effectiveness of the experienced worker, supervisor, and manager on tape talking about job hazards and assisting with orientation is considerably more effective than the use of professionals. Professionals can be used in the direction, sound, and other technical aspects of video making, but not as actors.

We have been discussing the disadvantages of using real actors in orientation videos. Now let's look at the disadvantages of using real workers in our videos. There are some disadvantages of using real people instead of actors. Some workers will be dissatisfied with their role in the video. Some people might not be chosen at all and might feel slighted. Some team members might be camera shy or not proud of their appearance. Some people may not speak clearly or loudly enough so that others cannot understand them on the tape. There may be language barriers or a failure to enunciate clearly that causes some potential team members not to be used in the video.

There are other disadvantages in using real team members. That is, if we have people portrayed in the video as key people or experts in their field and these team members are no longer with the company, this can be a negative reflection on the company. It we have an orientation video starring workers who are no longer with the company, it probably means that the video is old and should be updated. We should be using videos that are of the highest quality and are modern. All too often we have all viewed old, outdated, poor-quality videos covering subjects that do not apply. Using this type of material is a poor reflection on our company and sends an unfavorable message to our new team members.

Using videos for a variety of tasks, including orientation, has become common practice in the recent past. In fact, if you have visited a large manufacturing facility or industrial plant recently, it is likely that before allowing your entrance to the work area, you and all of the individuals in your party would have been required to watch a video. Besides watching the video, you would probably be asked to successfully answer some questions on a written test and not be allowed to pass until you have completed it.

Some facilities have many tiers of orientation. In these cases, the type of orientation that each entrant would receive depends on a variety of factors. The orientation could depend on the areas of the plant you will be visiting, the amount of time you will be spending there, and the type of work you will be doing, or other factors.

Although this system may seem complicated, it can work well in a large facility that does an appreciable amount of work through outside contractors. This system is a tiered approach. Depending on what type of work is being performed, the orientation will change to match it. If the work is short, simplistic, and unlikely to adversely affect the operation of the plant, the orientation for those workers can be short and simple. If the work is scheduled to be lengthy, the orientation can be lengthy, but the anticipated length of time that the work is scheduled to last is not the limiting factor. The limiting factor is more the likelihood that workers will get into trouble from a safety and health point of view.

When numerous videos are used for orientation, someone needs to use judgment to determine which contractors need to view which orientation video. Whenever individual judgment is used, it leaves the door open for criticism or mistakes. Not that anyone would ever cheat, but someone needs to ensure that the contractors are actually viewing the orientation videos. If tests or quizzes are given as part of the orientation, someone needs to ensure that no one is cheating.

The type of work that is scheduled to be performed and the general area where the work will take place is more important than the length of the job in determining which orientation video (or videos) is presented. One needs to carefully consider the following questions before conducting orientation or allowing outside contractors to work in any facility:

- Do the conditions in the work area warrant that workers be required to wear hearing protection?
- Do the conditions in the work area warrant workers to wear other personal protective equipment (PPE)?
- Does the prospective work area have off-limits sections?
- Are there any other training requirements for persons or outside contractors to enter a particular department?
- Does the prospective work area have moving assembly lines, heavy equipment operation, spray painting, quenching, ongoing construction activity hazard, or any other hazards that the contractor employee needs to be cognizant of?
- Are there security considerations?
- Is constant supervision or escort service required?

Example: Orientation for an On-site Visit

Let's look at the specific example of a contractor employee visiting a powerhouse to bid on some work. What type of orientation does this contractor employee need in order to perform his work? Even though this scenario appears to be basic, before answering the question let's ask ourselves a few questions about the sensitivity of the facility:

- What are the safety, security, company-secret issues in the area that you need to visit?
- Is the powerhouse a nuclear facility?
- Is there special training that one must have to reach the area that needs to be visited?

Once we answer these questions, it is likely that a lot more questions will arise, such as:

- Will the contractor employee need to physically touch or measure anything to complete his bid?
- Is the contractor familiar with the plant area?

(continued)

- If the contractor gets lost, can he or she wander into a restricted or dangerous area?
- Are there other contractors or other work going on in the area that the visiting contractor employee should be cognizant of?
- Will the contractor be required to wear a respirator or any type of PPE while touring the facility? If so, will the contractor be required to show medical fitness to wear the PPE?

At first glance, a situation that looks simple and straightforward may contain many twists and turns. Failure to properly orient contractors or in-house team members can be a serious error. The use of videos as an orientation tool needs to be carefully considered on a case-by-case basis.

THE USE OF INTERACTIVE COMPUTER ORIENTATION

This being the Information Age and the age of the computer, you might think that some type of interactive computer orientation would be available to you. Well, it probably is. Many companies in business today will customize not only orientation but also training in the form of an interactive computer program.

It seems like we are increasingly moving in the direction of computer interactivity and away from human contact. Computers—like videos—can be a great tool, and interactive computer-based courses have been shown to disseminate much information to the user; however, as in the case of videos, I am becoming concerned about us getting too far removed from human contact.

Although there are glitches in every system, it seems there are always glitches in computer interactive training. Even without computer glitches, there is one disadvantage in using computers that seems obvious. That is, many of the people whom we are trying to orient may not be computer literate. So, even if you have developed great interactive programs, you may need to train your participants on the operation of a computer or offer some alternatives. If your firm has not quite yet reached the interactive computer program age, you might consider the use of a videotape backup.

But let's step back a bit. If we use interactive computer programs, videos, overheads, and written material to assist in new hire orientation, one might ask why we need to have any humans involved. After all, this is the electronic age. Why should we try to deal with humans for orientation when we can do the whole thing electronically? Humans need to be involved in all phases of the orientation, including involvement, or at least availability and follow-up, during computer interactive orientation. It also means that a human should be involved to reinforce important concepts and point out areas in the film or computer program that may not apply to that facility, are outdated, or are in need of specific site information to be useful.

Human involvement is the key. And not just any human involvement, but someone who knows the company well and can communicate effectively and accurately at all levels. If the new hire needs specialized training before he or she can start work, the same principles are recommended.

Again, let's recall the Hawthorne Effect. Consider taking the typical widget department worker and using this worker in an orientation video. In our scenario, the expert is an experienced worker who shares his expertise with others on videotape.

Other people might believe that the widget operator is a lower-echelon, nonessential cog, but not us. We know that this worker is an important team member providing an important service to our organization. This worker's expertise is important to all of us and to the survival of our team. We need the widget operators on steps 5 through 8 to show up every day and produce quality widgets that are better, safer, cheaper, and faster than our competitor's widgets. In fact, we need all of our people to show up every day and perform their work tasks productively with quality, safety, and all of our company's core values in mind.

SUMMARY

Orientations are an important part of any new hire's career. This important step can start one off on firm footing or shaky ground. Everyone needs to be oriented. This includes the janitor through the CEO. Safety needs to be stressed during the orientation. Some very important information needs to be covered as soon as possible. Some of this very important information includes how does one contact the police, fire, or emergency responders, along with where does one go, or what is one's role during an emergency. Orientation should match the work duties and responsibility. The janitor's orientation would likely stress the locations of certain items and not video conferencing whereas the CEO orientation might stress phone or video conferencing, etc.

Videos and computer interactive programs are being used more and more in recent times and can be an important tool. Keep in mind that both videos and computer interactive programs have certain drawbacks and should not be used as a substitute for human interaction.

7
Training

Training is an important part of safety culture development. The worker who is adequately trained will be a safer and more productive worker. Building and developing safety culture is a step-by-step process. Training is one of the most important steps in the building, development, and enhancement of safety culture. Let's go back a few chapters and take a look at the various steps that lead us to the training step.

We started with some basic steps. Going back to the information covered in previous chapters, we took the time to screen and hire the best available applicants. We ensured that they had the physical capabilities to perform their work tasks by sending them for a physical examination. We ensured that they were mentally prepared or at least tested them for the influence of alcohol and drugs. We gave the team member an extensive orientation, both office and field. After thorough orientation, we have finally reached the training step. With appropriate training we ensure that team members are prepared to properly and safely perform their assigned work tasks.

The training that the team members will receive may be performed on the company premises, at a local educational institution, at a training facility, or just about anywhere we can imagine. Where the training is held is not the main issue. The primary issue is the *quality* of the training.

HOW TRAINING AFFECTS SAFETY CULTURE

No team member should be expected to perform any part of a job unless that team member is physically and mentally able to perform the task and has been properly prepared through adequate training. So, how does one ensure that team members are adequately trained? Craftsperson programs in which a worker has completed an extensive apprenticeship under the guidance of a skilled craftsperson and has successfully completed written and practical testing and a certification process would be one way to ensure adequate training. This certainly sounds logical, but it may not reflect reality. Many skilled craftspeople may be qualified but are not adequately trained or for a variety of reasons do not work in a safe or efficient manner.

Example: All Workers Need Training

> A skilled maintenance mechanic is hired who has previous experience from a competitor. One of the assembly lines breaks down, and this new mechanic is assigned to fix the problem. Even though this new mechanic is experienced, he or she probably does not have the experience with this particular assembly line machine.
>
> Although this new mechanic may have the basic qualifications to perform the work task, he or she is lacking the history of this machine and is not familiar with the machine's operation. Throwing the new mechanic into an unfamiliar situation and expecting him or her to fix the problem without an adequate background is similar to throwing a nonswimmer out of a boat in the hopes that he or she will learn to swim quickly—or drown. It is probably a better scenario that someone who has worked on this machine in the past and is familiar with its repair history and idiosyncrasies will work with this new mechanic for awhile. This mechanic will probably also be enrolled in some courses that educate attendees on the important maintenance aspects of this line.
>
> Let's look at the example of an auto mechanic. One would think that when a mechanic leaves one dealership and goes to work for a competitor, he or she can wheel his or her toolbox right in and get started immediately. After all, this mechanic is a skilled worker and has worked on these exact same cars for many years—only the location has changed, right? Not really. Although a team member's skill

and ability in diagnosing mechanical failure will transfer from one employer to the other, many items do not transfer. There are usually many different aspects for the mechanic to learn, for instance, where to get parts and dispose of used oil, uniform policies, tool policies, and the computer system or idiosyncrasies with car lifts or diagnostic equipment. Even the skilled mechanic needs to learn many new things at a new job. Just because someone holds craftsperson status or has a particular license does not mean that no more training is required.

Let's look at another example: Many drivers pass simulator testing, driver's education, and state-approved driving tests and obtain a driver's license. Most people would agree that even though certain people have obtained their driver's licenses, that does not necessarily mean that they can drive safely and efficiently. Even though the well-established trucking companies typically hire only those qualified licensed drivers with many years of driving experience, they will undoubtedly put newly hired drivers through their own training course, even if that driver is a seasoned veteran.

Let's consider a professional-level person who may hold a certificate or degree. In my experience (and probably yours, too), many people who are qualified (at least on paper) to perform a certain task fall woefully short of the mark when it comes time to execute a given task. Even though we generally hire only those persons who are believed to be adequately trained and qualified workers (who we believe are working in a safe manner), we may find that this is not always the case. Having the proper credentials, appropriate education, certifications, licenses, or the appearance of being otherwise highly qualified may be a good start but is not a final solution to any situation.

THE JOB SAFETY ANALYSIS

So, how does one ensure that adequately trained workers are performing their assigned tasks safely? There are no guarantees, but a good step is to have a qualified person perform a job safety analysis. The job safety analysis (JSA) has taken on many names over recent years. Some people refer to safety task analysis (STA), job hazard analysis (JHA), or other names. In this book, this useful tool is simply referred to as a JSA.

Simply stated, a JSA breaks down work tasks into key components. This may start with prejob tasks that take place before the inception of work activity and continue through activity completion. Each task component is broken down into steps, and each step is analyzed for potential hazards. Once the hazards from each step are identified, control measures or a protective system(s) is chosen that protects the worker performing the task from the hazard(s). These protective systems typically include engineering and/or administrative controls and/or personal protective equipment (PPE).

The Occupational Safety and Health Administration (OSHA) has outlined this procedure in a publication entitled *Job Hazard Analysis*. The main body of this OSHA publication is included in Appendix A of this book. The complete publication is available on the Web at www.osha.gov/publications/osha3071.pdf.

Table 7–1 shows selected components taken from a JSA that was developed for the task of clearing and grubbing.

The amount of detail within the JSA can be an indication of safety culture development. Table 7–1 is an example of the type of information that a JSA can contain. This section highlights certain aspects of a JSA that was developed for clearing and grubbing. The original JSA for this task was many pages long. During the clearing and grubbing process, the worker may find that he or she must handle heavy objects or materials. This particular section of the JSA deals specifically with this small part of the clearing and grubbing task.

Table 7–1
Selected Components of a JSA

Potential Hazard	Hazard Control Measure	PPE
Handling heavy objects	• Successfully complete back safety training. • Observe proper lifting techniques. • Obey the company's lifting limits. • Use mechanical lifting equipment (hand carts, trucks) to move large, awkward loads. • Get help when moving heavy objects.	• Wear a back belt (if required). • Use work gloves.

The use of a table coupled with a bulleted format is encouraged. Other forms of JSAs can also be used; however, I use this format most often and prefer it to others I have tried. The bullets present this important information in the most concise and comprehensible way. Each key point is separated, and not too many words are used for the descriptions. This keeps the JSAs relatively simple, even though they may be lengthy.

Looking a bit closer at Table 7–1, we see that several areas could easily be expanded. Looking at the Hazard Control Measure column, notice that the use of mechanical lifting equipment is mentioned. If the use of lifting equipment was truly anticipated, another component would need to be included that analyzed the hazards involved with using trucks, hand carts, and so on. This added component (trucks and hand carts) could add tremendously more information to this JSA. Although the use of a hand cart is a relatively simple addition, the use of a truck can be more complicated because it could involve different hazards, more procedures, different training requirements, licensing, and so on. Potential anticipated hazards should also be analyzed in the JSA.

Notice that one of the Hazard Control Measures listed in Table 7–1 included required training. Some people prefer to have training requirements listed as a separate column similar to PPE. Listing training in its own column in the JSA can be helpful when the task at hand is hazardous and the required training is extensive and varied.

Once a JSA has been properly completed, it becomes a valuable tool. This tool can be used to train new workers on how to properly and safely perform a new task or a task with which they are relatively unfamiliar. Or the JSA may be used to determine if experienced workers are performing their assigned task safely and efficiently.

Using a JSA as a training tool is important because of its relevance to the task at hand. Because the JSA was developed by analyzing and observing the task, the information within the JSA should be immediately relevant to that task when compared to a textbook situation. The terms used in a JSA should be easily understood by the worker, again keeping the JSA relevant to the task at hand. These two concepts, *relevance* and *understandability*, add to the effectiveness of your training and build the safety culture.

When can you have too much training? Training is like communication. One can almost never have too much communication, and one can similarly almost never have too much training. In the "real" world, however, only certain time and funds are allotted for training or communication. Therefore, because the money, time, effort, and resources that you spend on training are limited, every effort must be made to ensure that whatever training is offered is effective and that the training offered reaps the maximum bang for the buck.

Example: Training and Safety in the Front Office

Let's take some specific examples of training and talk about their effects on safety culture. Let's look at the receptionist or front office person. The receptionist, although unlikely to add considerably to accident statistics, can add to a company's safety culture. The receptionist needs a safety orientation just as much as a factory worker. Let's look at the type of safety-related information the receptionist or front office contact person needs to know and disseminate:

- The receptionist needs to be able to recognize an emergency situation and take appropriate action. A typical situation may be using a fire extinguisher to snuff out a small wastebasket with burning paper or recognizing that phones do not work or that the power, heat, or other utilities are not working. In each case, the receptionist should know what to do and whom to contact. In some cases, the receptionist needs to page someone in the building or contact that person via cell phone.
- The receptionist needs to know where the stairwells are (in a highrise situation) and what the evacuation routes and rally points are.
- The receptionist needs to know how to handle answering the phones during evacuation drills so that callers are aware of the drill but are not misled or alarmed.
- The receptionist needs to know what to do if someone comes in looking to rob someone or steal equipment or belongings.
- The receptionist needs to know the location of the first-aid kit and fire extinguisher.

All of this type of information is safety training/orientation. If the receptionist does not know these things, management should consider taking another look at their importance.

Receptionists and administrative help in general have been the target of job cuts with the recent downturn in the economy. More and more when we call a business, we will find that no one answers the phone. Instead, phones are being answered by automated answering systems. The technology behind some of these systems is amazing, but when we replace our receptionists with a machine, we

> have also affected our safety culture. If your business has permanently lost a receptionist to an automated system in the recent past, you need to consider who will pick up the safety-related duties that the receptionist was responsible for.

THE BASIC STEPS OF TRAINING

Having team members who are adequately trained to perform their work tasks is a priority item for all firms. We need to make the training as good as it can be. Training objectives must be determined before initiating any training. In other words, someone needs to determine what your people are supposed to gain during training, or what "level" your people should attain during training. The different levels are discussed later in this section.

Training can and should reflect field "reality." We cannot just "talk the talk" during training, or we have not accomplished what we set out to do. Training needs to reflect reality and not utopia. If the auditing of field operations has uncovered certain weak spots, then these weak spots should be mentioned. If a plan exists to shore up weak spots (and I hope it does), this plan should be mentioned. If we have had accidents, we should discuss our findings of the cause of the accidents and our plan of how we will improve. Unless your company has achieved zero accidents, I don't believe that any company can look at its safety (or other) record and say that there is no need for improvement. The way in which the company handles continuous improvement should be reflected in the training.

The basics steps for the training include:

- Determine the training objectives.
- Determine the who, what, where, and how of the training.
- Determine if the training met the objectives.
- Review, assess, and adjust the training.

Determining the Training Objectives

Let's talk about determining the training objectives. In general, we can go back to the Hawthorne Effect. As we train our team members, they will be assured that management cares about them and that they are important. So training should be a good thing—and most of the time it is; however, there are times when training can send mixed signals. Let me give a few examples of how the anticipated positive effect from training backfired.

Example: Too Much of a Good Thing

> I was involved at a large hazardous waste site cleanup in the Midwest that took many years to complete. At this particular site, an edict was made stating that anyone and everyone who entered the exclusion zone was going to successfully complete confined-space training and rescue. So, one of the training objectives at this site was to ensure that all team members entering the exclusion zone would have successfully completed confined-space and rescue training.
>
> At face value, this seemed to be a good idea. To provide a little background, the management team had grown tired of audit findings that pointed out training deficiencies. It seemed that the audit team could usually find people who had not completed required training to perform certain activities. To alleviate this situation, management had reverted to universal training requirements. These requirements came down to everyone needing to complete many types of training that would probably never be used. Management was willing to overkill on training because they believed that training is a good thing. (And basically, I agree.)
>
> On the other hand, the site would find that large crews were delayed because one or more key persons did not have the adequate training or certifications. More than a few times, various phases of projects were unavoidably delayed for long periods while large crews waited for one or two persons who possessed adequate, up-to-date training credentials. While the wait was going on for these trained people, large crews and large pieces of equipment were idle.
>
> The site was so large and the team members moved between departments so rapidly that it was difficult to keep crews trained as needed. But management did not *ever* want to hold up work because of lacking adequate training credentials. From an outsider's point of view, it would appear that there should be an easier way to solve the dilemma. People pointed out that with a little more organizational skill, the problems could be resolved; however, the problems

were not resolved, and the universal training edicts became commonplace.

It turns out that truck drivers had to enter the exclusion zone; therefore, they had to complete confined-space training and rescue just like other team members who entered the exclusion zone. The truck drivers protested not wanting to spend time seated in a training class all day taking training that they had no interest in and would never use. The protests fell on deaf ears. The edict had been made and would be followed!

Early on in the confined-space and rescue class, a rather portly truck driver had been assigned to complete training. One of the exercises for mock rescue included pretending that a team member had been disabled while performing a work task in a confined space. The hole watch person would notify the rescue people, who had already set up a tripod that was sitting over a mock circular entry. The (mock) disabled truck driver was already equipped with a safety harness and lifeline attached to his body and a rescue tripod and winch.

All that the rescue team had to do was to crank the winch and lift the (mock) disabled truck driver through a reinforced cardboard tube through the opening to "safety." Even with all of these precautions, things did not go too well. As the rescue crew cranked the winch, the truck driver's body was being lifted through the tube and toward the opening. But as soon as the truck driver's midsection entered the tube, he advised the rescue crew that his body was stuck in the tube.

At first, the rescue crew was unsure that the truck driver's body was truly stuck, and everybody got a good laugh out of the truck driver's warnings. They kept cranking the winch and continued trying to pull the truck driver through the tube. Unfortunately, the truck driver was stuck, and the winch was forcing this man's body to be pulled through the tube. This incident resulted in an avoidable injury. The person who got stuck in the tube suffered pain and humiliation that became known and talked about throughout the site very quickly.

I mentioned that this accident was avoidable—and it was; however, I also believe that almost all accidents are avoidable. But getting back to training, it turned out that other visitors and certain other team members went through the exclusion zone without being required to have confined-space and rescue training. Why was it so important that truck drivers be trained in confined-space training and rescue?

(continued)

> The answer is that sending team members to universal training that there is no regulatory driver for them to attend and that has no or a very limited useful purpose should be avoided whenever possible. Forcing team members to attend training that they have no real reason to attend (except for the in-house edict) should be reconsidered. This approach casts a negative light on the development of safety culture and casts aspersions on training programs. Although the training objective stated at the beginning of this example seemed to be a good idea, the end result did not turn out that way. The management at this site should have looked for other ways to ensure that projects were not unduly delayed because of a lack of properly trained workers.

Training should make sense. Although some people believe that regulatory drivers force management to have team members complete training for less than worthwhile purposes, I do not agree. Training can and should be interesting and useful under all circumstances. Some guidelines are offered later in this chapter as to how to accomplish this goal.

Example: Going Around the System

> One of my previous employers had a policy that no one would be allowed to have their HAZWOPER annual refresher training expire. The interpretation for expiration was that the time frame between refreshers could not exceed 13 months. As in the previous example of mandatory confined-space and rescue training, this seems to be an admirable goal.
>
> A computer-generated reminder system was in place. Each individual who was enrolled in the program would get a reminder at the beginning of the 11th month that their annual refresher training was coming due. The reminder would come again at the end of the 12th month. On the 13th month, a red flag would go up. A notice would go out to both the delinquent team member and the business line manager of the delinquent team member advising them that training was overdue. The notice would further state that arrangements must be made immediately to have the team member HAZWOPER refresher trained.
>
> This system often translated into taking extra time and effort to accommodate those who were delinquent. Depending on who you

were, and how well you got along with your boss, it might mean that you would be given a plane trip to some other city where an in-house HAZWOPER refresher training was taking place.

Most people, when they received their first notice, were diligent to ensure that they could schedule their refresher course and complete it within an acceptable time frame, but there always seemed to be some mid-level managers who were headed for an extra plane trip every year. This extra plane trip usually took three days to complete rather than the eight hours the course should have taken when properly scheduled.

It appeared to many people that if you missed your training class, you could actually be rewarded with an all-expenses-paid trip and get away from it all for a few days. As you can imagine, this sent a mixed message to the staff. Why is it that if you are not diligent in scheduling your refresher course you get sent for a three-day training trip? This situation repeated itself several times and did not sit well with the staff.

This situation should be addressed during performance reviews. Besides being time consuming, sending anyone off for this type of training is wastefully expensive. Last-minute plane trips, rental cars, and lodging are expensive. Besides the time and expense, we have sent mixed signals to our team members. Why should one of our members be rewarded with a trip for being delinquent? They should not. And when a team member receives a free plane trip for being delinquent, we have taken a step in the wrong direction as far as developing safety culture.

Instead of approving a trip for the team member, we could consider contracting a local vendor to provide the training, or possibly delaying field activity for this person until the course is completed locally. There are numerous other ways to solve this situation, but we do not want to provide negative reinforcement.

The previous two examples demonstrated how good intentions for safety training had produced undesirable results. Let's move on from what not to do and talk about what should be done. The first thing we should do is to determine the training objectives.

Determining training objectives can be a difficult task. We find that there are three basic drivers to training:

- Mandatory training required by a regulatory agency
- Mandatory training required by in-house guidelines

- Training not mandated by in-house or regulatory agencies but that is still important

In the previous two examples, both types of training stemmed from regulatory agency requirements. For the confined-space and rescue example, the likelihood for the truck driver to actually utilize this training or need it was minute. For the HAZWOPER refresher training, it was likely that this team member would need to have completed the course to stay within regulatory compliance. Even though the policy focused on compliance, to send a team member on a plane trip to complete a required training when it could have been done in a much more efficient manner is a situation in need of correction.

So, looking at the three types of training, an example of the first type, regulatory-agency required training, would be our HAZWOPER refresher example. For the second type, training required by in-house guidelines, confined-space and rescue is a good example. The third type of training, which is not mandated by regulatory agencies or in-house requirements but is still important, has not yet been discussed. To demonstrate this type of training, consider front-line supervisor safety training. I believe that having properly trained front-line supervisors strengthens and develops any company's safety culture.

Let's discuss training objectives, keeping the three types of drivers mentioned in mind. For mandatory training as required by regulatory agencies, the driver for this type of training is regulation driven, and these regulations are law. If we want to be law-abiding citizens and good corporate citizens, we will abide by the law and obey the regulations. So, one of the training objectives for *regulation-driven training* is to obey the law.

After ensuring that the training is compliant, we can focus on the next step, which is to make the training relevant. When we examine HAZWOPER or any other training, there is an underlying reason why our lawmakers enacted that particular legislation, and it is probably related to our business activity.

So, the objectives of our training are growing. We want our training to be compliant and relevant. To be compliant, we must first know the law and understand how it pertains to our business. After we know the law and understand its pertinence, we must design our training accordingly. The compliance part should form a rough outline of our training course content. After our outline is developed, we need to fill in the areas with information that is relevant to our business.

Let's take the example of HAZWOPER training and examine the training objectives. For a company that is in the business of removing underground storage tanks (USTs), the HAZWOPER training would contain information as specified by OSHA. Some of the general subjects covered might include the following:

- PPE such as safety glasses, boots, Tyvek suits, gloves, etc.
- Lockout/tagout
- Excavation/trenching
- Respiratory protection
- Bloodborne pathogens
- Hearing conservation
- Safe driving

This type of training program would probably be both compliant and relevant for a company removing USTs.

If the USTs contained only gasoline and diesel fuel, the agenda should include an in-depth discussion of the physical and health hazards of gasoline and diesel fuel. If the USTs contained other chemicals, those chemicals should also be discussed. The point is that the training should be *relevant*. Team members are generally required to attend this training annually. Because the business has not changed and the workforce has not changed, why should the format of our training class change? Sometimes in a mature business such as underground tank removal, the team members attending annual training are doing so every year for 10 years or more. It is certainly a challenge to keep the class compliant, relevant, and interesting for 10 or more years using the same basic material.

This certainly is a challenge, but the challenge is not insurmountable. If we look at something as mundane as safe driving, even this subject can become interesting and relevant over a 10-year period. There are volumes of different materials and videos that discuss different techniques of safe driving. The last course that I took used a 15-minute video that concentrated on driving safely in city traffic under crowded conditions. The video was well done, and although I have been driving in crowded city traffic for many years, and did not believe that I could learn any new techniques, I was pleasantly surprised. More than 80 percent of the class was pleasantly surprised with this video as well.

It is a step in the wrong direction to provide annual training and have the training become stagnant. If training is going to be performed, it needs to be done in a professional manner. As the experience levels of the audience change, the training must also change to keep their interest. The training should be constantly evolving. As the workforce and the work changes, the training also needs to change.

Training participants need to know what the training objectives are. If participants are going to be tested at the end of the course, they should be advised of this expectation. In the beginning of the course, the instructor(s) must mention what these objectives are. Let's examine the previously listed bulleted items for HAZWOPER training objectives.

The first bulleted item was PPE such as safety glasses, boots, Tyvek suits, gloves, and so on. It appears that the training objective would be to ensure that team members are expected to know what type of PPE they will probably use on the job during their work activities and how to use the PPE properly. If this is the case, the instructor should state this as a training objective and then include two to six questions about this subject for the test at the end of the course.

The second bulleted item was lockout/tagout. The training objective might be for team members to refamiliarize themselves or become familiar with the lockout/tagout program for the company and know when and how to use it. If this is the case, the instructor should state this as a training objective and then include two to six questions about this subject for the test at the end of the course.

The third bulleted item was excavation/trenching. The training objective might be for team members to know general rules about excavation/trenching so that they can determine when a competent person is required on site, among other things. Typically, the training offered in a refresher course might not qualify the attendee to be a competent person for excavation/trenching, but should provide more of the "survey" information to give a person a "taste" of the subject rather than in-depth knowledge. If this is the case, the instructor should state this as a training objective and then include two to six questions about this subject for the test at the end of the course.

The previous point is very important. Some people have walked away from training classes believing that they are qualified to be competent in a variety of areas. The training of a competent person in the area of excavation/trenching has been confusing to some participants. This point must be made clear and emphasized. No one should come away from a training course misunderstanding what that course has qualified them to do. *Note*: To make this point unmistakably clear, consider specifically adding one of the two to six questions pertinent to this section.

The next bulleted item was respiratory protection. While removing USTs, there may be a potential to use a respirator. The training objective might be to determine under what circumstances a respirator would be used and possibly how to use it. If the instructor plans on fit testing the attendees, this should be made clear before the beginning of the class so that the participants will have their respirators with them. If this is the case, the instructor should state this as a training objective and then include two to six questions about this subject for the test at the end of the course.

The next bulleted item was bloodborne pathogens. The training objective might be to cover the company's bloodborne pathogen policy and how it relates to UST work. If this is the case, the instructor should state this as a training objective and then include two to six questions about this subject for the test at the end of the course.

The next bulleted item was hearing conservation. The training objective might be to determine under what circumstances hearing PPE would be used and possibly how to use it. Another training objective might be to cover the company's hearing conservation policy and how it relates to UST work. If this is the case, the instructor should state this as a training objective and then include two to six questions about this subject for the test at the end of the course.

The next bulleted item was safe driving. The training objective might be to learn two techniques to aid while driving in heavy rush hour traffic in a large metropolitan area. If this is the case, the instructor should state this as a training objective and then include two to six questions about this subject for the test at the end of the course.

Determining the Who, What, Where, and How of the Training

So far we have determined what we want to do with our training. We now need to determine the who, what, where, and how of the training. Let's talk about the "who" first. If the company is large enough to have a training person, course participants have probably seen this training person perform on several occasions. As previously mentioned, it becomes a challenging experience when we have the same audience reviewing the same basic material for year after year of annual refresher training.

If your company is in this situation, you need to consider making some adjustments. There are usually ways to make even minor adjustments so that even if the basic information is changed, the viewpoint or approach can be different. If we look at the HAZWOPER content cited in the previous section, we see at least six different modules. Your company training person can be more of a course coordinator and bring in some guest speakers for different segments or modules of the training.

For instance, if the subject is respiratory protection, it is usually not difficult to find a manufacturer's representative to provide training on this subject. If you have a section on instrumentation, a manufacturer's representative might also be considered for this module. Some type of rotation should be used so that at least most of the material does not get stale or old. You could also consider changing the way in which the material is being presented rather than a change in the material. For instance, if your training is dealing with USTs that once contained gasoline or diesel fuel, consider putting together a crossword puzzle and presenting the information in that manner rather than via straight lecture.

If a course is more than two hours, consider using at least one additional instructor. I believe that the best length of time for instructors and students is about one hour. After one hour, a short break is suggested. If it is possible, courses should have at least one instructor for each hour of class. If the course lasted eight hours, you should have eight different instructors. Think back to your high school days. At the high school I attended, I would typically get six classes per day that were approximately one hour each in length with six different teachers per day.

In the real world, due to budget and logistical constraints, this rarely occurs. In fact, many excellent courses lasting eight hours have only one person as an instructor for the entire length of the class. But even if the instructor is fantastic, I believe it is an improvement to rotate instructors every hour or two, or as often as possible.

If your company uses an outside source for training services, the "who" part of this equation is a relative unknown. Consider doing as much in-house training as possible and keeping this training as relevant to the work as possible. Sometimes occasions, however, will warrant using an outside source.

Let's consider the annual HAZWOPER training again. For this training, a regulatory driver mandates that certain team members must have this training while performing certain activities or be out of compliance. Let's say we hire a new employee who has been out of work and that his refresher training needs to be renewed. And let's say that we hired this person to perform work on a project that is starting in three days. We find out that our in-house training cannot happen within the time frame for whatever reason and determine that the best way to accomplish the training is through an outside vendor.

As I mentioned earlier, my philosophy is to use as much in-house relevant training as possible, but there are circumstances when it is considerably simpler to use an outside source. I must stress, however, that it appears that with a little more planning, the aforementioned situation would not have happened. Hiring managers should work closely with those persons responsible for training so that these types of situations can be avoided. But let's say that even with all of the planning and communication we can muster, the situation with someone needing training ASAP occurs. We determine that the best way to accomplish this is through outside vendors. When this becomes the case, we now have a new "who."

This "who," along with the "where and how," needs to be considered. All this takes upfront planning. A determination must be made of which vendors are available, who they use for instructors, where they perform the training, and how they will accomplish the training objectives we have determined.

Answering *who* will be performing the instruction is certainly an important consideration. Even if we have great instructional material, a wonderful classroom atmosphere, and an attentive class, an instructor who is less than highly qualified or lacks professionalism or interest will detract from this important experience.

Starting with a competent instructor is a good first step. After locating our competent instructor, consider the addition of some charisma and the ability to read and involve the audience. After that, consider the addition of a relevant curriculum, including instructional materials that are relevant and up to date, with readable copy, and generally well done. We couple this whole package with pleasant classroom conditions that are conducive to learning, good visual material (e.g., overheads, PowerPoint presentations). If everything goes according to plan, this may turn out to be a positive learning experience—and one that will help develop our safety culture. If shortcomings are noted in any area, adjustments for improvement should be made.

Although "where and how" were touched on briefly, let me go over these two areas in a bit more detail. The *where* can be important. Advantages and disadvantages can be attached whether one chooses an in-house location or an off-site location for training. Some locations (whether in-house or off-site) are not appropriate for training. A couple of important considerations include the general atmosphere and logistics for appropriate training facilities. Strive for an atmosphere that is conducive to learning. But while the atmosphere is always important, do not underestimate the importance of timing.

For instance, while I was responsible for training at a previous employer, the need arose to train approximately 25 team members. The conference room sat 10 comfortably and had poor ventilation. In the past, the lunchroom was used for training. The lunchroom had some fine attributes for a training facility. It was centrally located, well lit, comfortable, well ventilated, and roomy. It also had some negatives attached to it. These included:

- Poor acoustics (noise seemed to reverberate from the ceramic tile floors)
- No real front/rear or seating arrangement
- Constant interruption from the vending machine area

- Constant interruption from the paging system
- Many different people just "wandering" through to see what was happening

In all, it appeared that the negatives outweighed the positives, but in this situation one positive outweighed all of the negatives—it was cheap! In fact, it was free. My job was to make this training as good as possible within the budget. In order for me to accomplish my mission, some adjustments needed to be made.

Described as follows are some of the adaptations that were made so that the training could be as good as possible under the circumstances. First, it was decided that the training would be held on a Saturday. This eliminated the negatives associated with the interruptions at this location. To eliminate the reverberation from the ceramic tile floors, some mats were placed to keep the noise down. To address a workable seating arrangement, the best angles were considered for lighting. Once the lighting was addressed, the tables and chairs were set up as if the lunchroom were a classroom.

This training session became the best training situation possible in view of the constraints that existed. Such constraints as previously listed are common when attempting to set up training. Adjustments have to be made, and allowances and compromise are typically in order.

Getting to know the outside vendors can be another struggle. We talked about the importance of the "who." The "who" is important for both in-house instructors and the instructors of outside vendors. Finding competent instructors from outside vendors can be a difficult task. To determine if the outside vendor's instructor is the person you would like to teach your team members, consider the following tasks:

- Interview the instructor.
- Ask for a video presentation.
- Request attendance at or at least to sit in on a small portion of a course that the prospective instructor is teaching in the near future.

Sometimes, depending on logistics, your choices of instructors may be limited. Just as in the example of using the lunchroom for a training room, sometimes you have to compromise to move forward. Whether the instructors work for your company or are independent, they should be qualified to teach the subject by means of work experience, training, licensing, or education. Besides being qualified, they should be good communicators. In order to determine just how effective your prospective instructor is, his or her communication skills should be evaluated. As mentioned previously, the most effective way to check on the communication skills of your prospective instructor is to actually take a class and witness the skills firsthand.

If you cannot witness the skills of the trainer firsthand, there are other ways to determine the communications skills of an instructor. One of those ways would be to pool comments from other companies that have used that particular instructor. This can usually be accomplished through a trade association or local professional society. Someone from the association or professional society can probably offer input regarding the quality of the instructor and training where a particular instructor taught.

It should be a relatively easy task to pool other members of a trade association or professional society or other group and determine who they use for training and how effective the training was that they obtained. If these other firms agree that the training they obtained was up to par, that would be one indication of the quality of training. After receiving recommendations from other companies, consider taking a ride or field trip to the training facility, meeting the proprietor, and getting a tour. One should be able to determine if the facility is serious about training when the classroom setup is observed. Some things to consider are:

- Do they have classrooms adequate to serve your needs?
- Do they have audiovisual equipment adequate to serve your needs?
- Is there adequate parking available or public transportation to the door?
- Is the location of the facility convenient to your participants?
- Does the facility appear to be clean, safe, and appealing?

There are probably a variety of other things to consider when looking at the facility. If the interest in training is only for instructors to come to a client facility to perform training, maybe the condition of the facility is unimportant. Maybe all of the training that you will be looking to perform will be completed by others but at your facility. If this is the case, the outside vendor's training facility may have little bearing on your choice.

Once you get past the "who," you can get to the "where" and "how." For the where and how, remember the previous discussion with the example of using the lunchroom as a training room. The where may mean that we want our outside vendor to perform the training at our premises. Or the where may mean a hotel suite or conference room, or just about anywhere we choose.

The how is also important. Many outside vendors use computer-based training. Some people are satisfied with interactive computer-based training. This type of training has some obvious positives associated with it. If you have field people who have little free or office time available, being able to take courses such as a HAZWOPER eight-hour refresher online can be an advantage. Using computer-based training can take the who, where, and when out of the equation because when can be anytime, where can be anywhere you have a phone line with a laptop hooked to it, and who is taken completely out of the picture.

There are also disadvantages with computer-based systems. Some of those disadvantages include:

- Lack of computer literacy
- Lack of computers
- Technical difficulties with local phone lines
- Software problems
- Lack of anyone to answer questions or provide further explanation

And this list could be expanded.

Determining if the Training Met the Objectives

Let's say that our training objectives have been determined. And that the *who*, *what*, *where*, and *how* of training have been determined; and, we have just completed our first training class. We are finished, right? Wrong.

Even though who, what, where, and how, along with successful completion of a training class, are behind us, we are not finished. We now need to determine if the training truly met the training objectives. In other words, did this training take us where we wanted to go? Ideally, the answer is a resounding YES; however, this needs to be definitively determined. Some useful techniques can be employed to determine if our objectives have been met.

One technique to determine if training met objectives is proficiency testing. If the attendees passed the test demonstrating adequate proficiency, then the training met objectives—or did it? The answer is maybe. Proficiency testing can be an effective and useful tool, but it is not a know-all, end-all. Another consideration is whether the knowledge gained in the training is used in the field. If our in-house inspections pointed out deficiencies and our training was supposed to address these deficiencies, then these deficiencies should no longer exist. In other words, management identified a problem and determined that training would resolve the problem. Once training is implemented, the problems should be resolved.

If at least part of the training objectives were to develop a course to address problems experienced in the field, and follow-up testing revealed continued problems in the field even though course graduates demonstrated proficiency on tests, then the problem still exists. Maybe the course objectives were not fully thought out. Maybe the questions need to be adjusted. Maybe the participants are cheating on the test. Maybe our culture accepts a wide gap between classroom knowledge and work output in the field. Or, maybe none of the above.

Let's say that a safety inspection was held during a confined-space entry procedure. During this inspection, the inspector noted that some of the air-monitoring instrumentation equipment seemed to have readings that were not inline with expected results. When the team members were questioned, the inspector felt that their answers did not reflect confidence in operating air-monitoring equipment. One of the recommendations from the inspection was that an air-monitoring instrumentation class be held so that team members can demonstrate with confidence their knowledge of how to properly operate air-monitoring equipment.

In this example, if the objective of our instrument training session was accomplished, then our inspectors should no longer observe improper air-monitoring instrumentation during future inspections. If inspections continue to reveal deficiencies in air-monitoring techniques, then the training needs to be adjusted.

Reviewing, Assessing, and Adjusting the Training

That leads us to our last point, which is review, assessment, and adjustment of the training. As mentioned in the previous example, if lapses continue to be observed in the field even after training has been successfully completed, then some adjustments are probably in order. Discovering in the field that team members do not understand how to properly operate air-monitoring equipment is a serious matter. Operation of this equipment in an improper or incomplete manner could result in serious injury or death to entrants or others in the area. One should never be able to observe this type of deficiency in the field. This lack indicates a failure in the system farther up the line. If this type of problem is observed in the field, that would suggest that an immediate, hard, and close look at not only the training program but also supervisory and management commitment may be in order.

It is hoped that problems such as those mentioned in the previous paragraph are never found in the field at your facility. Instead, skills such as air-monitoring instrumentation are handled at the classroom and laboratory level. If team members are going to be expected to operate air-monitoring equipment, then they need hands-on experience to do so before expecting them to perform in the field. The training can include workbooks, operation manuals, overheads, videos, classroom discussion, and so on; however, no matter how much book learning someone has, there is usually the need for hands-on experience to ensure proper understanding. In a situation such as air monitoring for a confined space, no one team member should be allowed in the field alone until he or she can demonstrate a competency level, confidence level, and proficiency level that shows no room for error.

So, if one were to review, assess, and adjust training, it would appear that initial confined-space training is in need of review, assessment, and adjustment. The program should be reviewed and assessed to ensure that air-monitoring instrumentation training is adequately addressed. If the assessment indicates the need for improvement, then adjustments in the program should be made as necessary. After making the necessary adjustments, we should again review and reassess training. If deficiencies are still noted, then more adjustments should be made. This process should be continuous so that programs are being assessed and adjusted accordingly on a regular basis.

Training should include testing. I believe that written tests should be part of any training. This would be especially true in a confined-space training scenario. I believe that it is most practical to have training tests be true/false or multiple choice. Other types of tests such as crossword-type puzzles, fill in the blank, essays, and other approaches can also be effective. But if they are properly designed, I prefer to use true/false and multiple-choice questions to assess training. The test should cover whatever subjects were supposed to be learned during the training.

Training objectives should be carefully examined (as discussed earlier in this chapter), and testing should be engineered to determine if those objectives were met. If we look at the field deficiency noted earlier, a test question should be added to specifically address that deficiency. If the test question is already part of the training, and workers answered the question properly on the test but still lack the skills in the field, there is more investigating to do.

Here are some potential reasons why a person would be able to answer a test question properly but act differently once they leave the classroom and enter the field. The team member may have just guessed right on the test, but really did not understand the topic. The team member may have been looking at a neighbor's answer sheet and copied that person's response. Although the first two points are a possibility, I think the main reason is that although the team member successfully answers questions in the classroom, the company culture and past practice accepts a gap between classroom and field reality. I have heard many times that something "looks good on paper." This indicates that the plan may be adequate but that there is some doubt that the plan will be executed as currently written.

Field adjustments are part of reality, but as we improve the safety culture, the gap between the field and the plan (or the textbook or the class) should be minimal. If our company's safety culture accepts gaps between plans and field reality, then we need to fill those gaps as soon as possible. In the case of air-monitoring instrumentation, there is little room for error. There should be no gap between the plan and field reality. The proficiency of the team member operating air-monitoring instrumentation should be at an adequately high level.

Testing is a way to indicate proficiency level, and determining a proficiency level for duties such as air monitoring should definitely be determined. If the information is so important, sensitive, or integral for the team member, then the proficiency level should reflect this fact.

Management needs to get input from those persons attending the class. If attendees have comments for class improvement, those comments should be carefully considered and acted on if appropriate. An evaluation form should be used for every training session that obtains the team members' assessment of training. These evaluation forms need to be reviewed and adjustments in the course should be made as required. Besides the team members who are currently taking the course, we should also gain input from those seasoned workers as to how we could make improvements to our training. If the team member gained certain information on the job rather than in the class, one should be cognizant of that. If the course curriculum can be adjusted to reflect suggested improvements, then it would improve our program to do so. Even if we do not get input from our current workers, a review of training programs should take place as needed, but no less than annually.

DRIVERS AND PROVIDERS OF TRAINING

In the beginning of this chapter, three drivers of training were mentioned: (1) regulatory required training, (2) in-house required training, and (3) nonmandatory but important training. As mentioned earlier, having a well-qualified supervisory staff is a key item in safety culture development. The HAZWOPER standard is one OSHA standard that requires specific supervisor training. OSHA does not mandate this type of specific supervisor training for most other industries. So, for the most part, the type of training that our front-line supervisors should be receiving is probably not specifically regulated by OSHA.

As far as in-house requirements for supervisors, some companies have stringent guidelines outlining supervisor requirements, but many do not. Supervisors can be chosen for many reasons. It is not uncommon for the supervisor to be chosen because he or she was the most efficient worker in a particular department. Or, if a person worked in a particular department for a long time, he or she might be chosen as a supervisor. Although not always the case, when methods such as seniority or efficiency are used to select a supervisor, the person chosen may not have all the tools or personality traits necessary to do the job as completely as management would like.

On the other hand, the supervisor who is promoted from within the ranks has a great advantage over the outside hire for a variety of reasons. Those reasons include knowledge of the department operations, the people, the company, and the work in general. I will not offer more detail regarding this subject except to say that management will typically accept those previously chosen by any method to be a supervisor rather than place them back into the hourly ranks.

If your company accepts the current makeup of the supervisory staff, there will probably be a need to train and groom these team members so that they can perform their jobs more effectively. With human nature being inherently resistant to change, this task can be challenging. It is important that the supervisory staff understands the vision of the company and actively participates and provides leadership in any changes that occur. Obviously, top management needs to embrace changes and communicate the importance of the changes to all team members.

The role of front-line supervisors in most companies in the development of safety culture is very important. The front-line supervisors need to be believers and supporters of change and be able to support these changes to the people who actually perform the hands-on work. Having front-line supervisors who are prepared for their responsibilities is a never-ending and difficult task.

Supervisors need training and grooming to be effective at their jobs. This difficult position requires tremendous people skills along with a variety of technical expertise. This technical expertise includes a large amount of information that is pertinent to safety culture development. Later in this book you will find an entire chapter dedicated to training the front-line supervisor so that safety culture development is promoted.

We need to remember that the front-line supervisor's job takes a lot of skill and know-how in order to be effective. I believe that one of the most important skills is people skill. Understanding the Hawthorne Effect and how to go about using this effect on a day-to-day basis is a great skill to acquire. But people skills are not something one is born with; these skills must be developed.

Both large and small companies typically have the desire to help develop their supervisory staff. Some large companies offer an in-house course for supervisors and safety managers. Smaller companies use a variety of outside vendors for this purpose. I have attended and taught both types of courses and am most impressed with the in-house courses versus those purchased from outside vendors. In-house courses have some advantages over outside vendor courses. With an in-house course, participants typically meet other management members who are attending the course who will share problems and experiences with each other and typically can relate well. These relationships can go far even years after completion of the course. I have not noticed this same development of camaraderie with outside vendor courses.

Teaming is an important concept. The smaller employers that I have worked with have tried to improve the teaming effort by sending two to four team members from the same location to the outside course at one time. Grouping participants in this manner allowed those participants to get to know each other and deepen their relationships. So, even if you are a smaller company, teaming can be accomplished at outside vendor courses.

Large companies can enjoy the economies of scale. As the large company grooms and trains supervisory staff, they probably have more seasoned veterans within their staff to mentor and train supervisors. The small company typically does not have that pool of seasoned management to fill in the gaps.

When the large company decides to provide a supervisor training course, they will probably have enough candidates for an adequate class size. Many large companies could probably find at least six candidates who would benefit from a course like this at any one time. For the smaller company, they may only be able to provide one or two candidates, and taking more than one or two supervisors away from the operation at any one time could prove devastating to daily functioning.

As an instructor, I have always appreciated a class size of at least 6 but not more than 30. For fewer than six students, I would cancel the class. After all, plenty of support, money, time, and effort gets expended while putting on an in-house class, and the money that is allotted for safety training must be spent wisely.

I have been particularly impressed with in-house courses that have a minimum duration of one week. The best in-house supervisory development course I attended lasted two full weeks. This particular class had been postponed twice due to lack of enrollment. When it was finally completed, I left with an increased understanding of OSHA safety requirements and company safety standards. I also had confidence in what my employer expected of me regarding safety performance, along with an understanding of what to do when I felt things were not going in the right direction.

I believe that these types of supervisory training courses provide the type of grooming that supervisors need to perform their jobs properly. I support them wholeheartedly and would recommend them for any company that has the size to support them and is willing to put in the effort to make the training "good." Although nothing in life is perfect, an effort needs to be made to make the training as good as it can be. Everyone associated with the training can tell when substantial effort has been expended to make a training class successful.

On the other hand, when the effort has been minimal or haphazard, this also is evident. If we look back to the Hawthorne Effect, if we provide team members with a poor-quality training experience, we have indirectly shown them that they are not important, and therefore management cannot expect much from them; however, if we show team members that management expects them to be well-trained because management realizes that they and their jobs are important, we will reap the benefits of the Hawthorne Effect.

SUMMARY

This chapter illustrated that training is a key to safety. The level of effort expended to develop training will play a key role in safety culture development. An important tool for safety training is the job safety analysis (JSA). JSAs have many different names and different acronyms, but no matter what you call them, they are used to identify on-the-job hazards and the safety systems that can nullify these hazards. Using JSAs as a training tool allows the team member to think in a way that analyzes each work task for potential hazards. Hazard recognition is an important initial step in hazard avoidance.

The importance of determining appropriate training objectives was discussed. Establishing a clear objective is a first step in planning a training session. Once this objective has been established, we focus on the plan for reaching our objective. To assist in reaching an objective, one should be able to answer a variety of questions, including the following:

- Who?
- What?
- Where?
- How?

Once the training objectives have been established, the "what" becomes our training course material or curriculum. The training objectives determine what the curriculum should be. The material or curriculum will vary as the objectives vary. The "who" is twofold. We need to determine who needs to be trained. That should be relatively easy. The next phase of the who is who will provide the training. This part of the who can be a little more difficult. The who could be an in-house trainer(s) or an outside vendor(s). The "where" can be anywhere we choose. Some valuable hints are offered to ensure that the where is conducive to training.

How the training will be accomplished is another important consideration. Computer-based interactive training programs are becoming increasingly popular. These programs certainly have some advantages, but I believe that nothing beats human interaction when the course is well orchestrated and you have the right human. Part of the orchestration includes videos that are current and well done and overheads that are clear, concise, accurate, colorful, and professional, along with an instructor who is competent and an excellent speaker.

After the training is complete, we need to determine if the objectives of the training have been met. Our inspection and audit program can help us review and assess our training. Depending on what our findings are from this assessment, we may need to adjust our training to address our findings. The value of written testing is mentioned several times in this chapter. I believe that the extensive use of written testing should be encouraged. Records of the completed tests should be taken. The questions on any tests should relate to the job as closely as possible. If it is observed that certain questions or sections of the training seem more challenging to participants, adjustments should be made to address uncertainty and raise proficiency levels as needed.

We talked about the "drivers" for training. These drivers include:

- Regulatory drivers
- In-house requirements
- Not required but strongly recommended due to obvious need

The training required by law or by internal requirements can be easily understood; however, the training that has no driver can be the most difficult to address. An example of the training that is typically not required is supervisor training. We expect supervisors to be management's liaisons for changes. Supervisors need to be extensively trained and groomed so that they can be effective in their jobs.

This extensive training for supervisors should include a healthy dose of safety training. We talk about the leadership that we expect our supervisors to display. We also discuss the needs for supervisor training in both large and smaller organizations. The main point is that we must make training as good as it can be. It is important to keep not only the course content and curriculum in mind, but also the instructors, training facility, and every aspect of who, what, where, and how. We also continue to mention the Hawthorne Effect. When we put forth the effort to develop a good training session, it shows. What this translates to is that management cares about their team members participating in a "good" training session. The point is that management cares about team members because the team members' jobs are important to management. Hence the team members' work will reflect the Hawthorne Effect.

8

Safety Performance as Part of Performance Evaluations

PERFORMANCE EVALUATIONS

Performance evaluations are a tremendously powerful management tool. How effectively your organization uses this tool can make a big difference in safety culture and the general development of overall company culture. We will talk about the effective use of this tool throughout this section, but first let's ask and then try to answer a few key questions:

- Is there a link between safety performance and performance evaluations?
- What part of performance evaluations should be related to safety?
- Should safety performance be linked to performance evaluations?

This chapter addresses these questions and more. Before continuing with this discussion, let's back up and define some terms. Let's first discuss the relationship between the performance evaluation and the salary review.

Notice that the term being used in the first question was *performance evaluation* and not a *salary review*. The distinction between the terms *performance evaluation* versus *salary review* is important. I would hope that every team member would have numerous performance evaluations over the course of a year, but for most team members, it is likely that they will have only one salary review.

I realize that I have been very general regarding the term *team member*. I am well aware that many people get salary adjustments or reviews several times during the year. But for this discussion, when I refer to *team member,* I am referring to the typical worker who has the most effect on accident statistics. I am not including those workers who are paid commission or who get most of their income from bonuses or are company owners or principals.

This definition does not include those types of workers because in my experience, upper-echelon management, owners, principals, or workers who depend on bonuses have little effect on accident statistics or safety performance. For purposes of this discussion, when using the terms *team members* or *workers,* I will be referring to the typical factory worker, construction worker, technician, warehouse worker, assembly line worker, delivery person, general laborer, equipment operator, foundry worker, or other typical American workers—basically, the group of people that usually are the categories of workers who are involved in on-the-job accidents.

When we look at salary adjustments versus performance, one typically gets salary adjustments very seldom when compared to changes in performance. Unfortunately, too many team members seldom or rarely get a performance evaluation. This point will be discussed in more detail later on, but remember that team members and supervisors should be discussing perceived performance and areas perceived to need improvement. This performance rating should be done in writing, formally at least twice per year, and on a less formal basis, and probably not in writing, several other times during the year.

Team members' performance typically changes frequently. For most working people, salaries may change once a year. Performance as measured by production may change up or down depending on lots of factors. From year to year, salaries will typically drift upward for most people, with many people experiencing more of a cost-of-living adjustment on a yearly basis. Union-negotiated contracts have typically included increases that would have included a cost of living "plus" adjustments. Whether your company is unionized or not, you may see salary levels increasing at a faster rate than profit margins.

I have often heard from co-workers that they have not received a formal salary review or performance appraisal in many years. Many of my management associates who deal exclusively with union team members believe that most union workers do not seem to expect or want a salary review. Some members of management believe that union contracts can be difficult to work with. I specifically mention unionized workers because I have had many dealings with members of management who believe (or at least make the excuse) that their hands are tied because of the strong union they must work through to get any job done. I have also dealt with unionized and other workers who had little regard for their supervisor's opinion regarding job performance whether it was safety related or not.

I am not trying to pick on unions, nor am I touting them. I certainly do not believe that because your company consists of a mostly unionized workforce that good management principles should fall by the wayside. Remember that the Western Electric Plant where the Hawthorne Study was conducted so many years ago was a union shop.

Although certain members of management may believe that unions, or environmental legislation, human rights laws, labor laws, OSHA, and a variety of other factors make it more difficult to operate, this is the reality of operating a business in this day and age. As time moves forward, it seems that business gets more competitive. It seems that in order to stay in business, more innovative ways need to be found to increase productivity or establish a competitive edge.

Having to spend less money on claims and accidents is another way to obtain that competitive edge. The development of a company's safety culture will result in less money spent on claims and accidents. The effective use of performance reviews is another tool for "teaming" and culture development. This big picture of "teaming" and building safety culture must remain in the forefront. If we can keep safety as a core value and couple that with the importance of our team members, we will improve our safety culture and culture in general just as was done in the Hawthorne Effect as was noted in the 1920s and 1930s.

Example: Safety and Performance

> I recently met a group of environmental technicians who perform rapid-response cleanups in the hazardous waste area. These workers at a minimum might typically be required to receive two general performance reviews and one salary review per year. The written performance reviews took place in March and September, with a salary review in December and any increases taking place in January. This would mean that this worker's supervisor would be formally assessing this worker's performance in March, September, and December of each year.
>
> In March, the supervisor would meet with the team member and discuss and rate the team member on general performance. Part of this rating would include a rating on safety performance. The safety rating should be one of the first items that is discussed. The worker must have a clear understanding of what the expectations are for
>
> *(continued)*

safety performance. If the supervisor believes that an improvement should be forthcoming, he must point out to the worker why he feels there are shortcomings and what can be done to correct any shortcomings. Expectations of what the team member can do in the short and long term to correct shortcomings, along with what the company can do to support the team member, should be discussed during performance evaluations.

Some of the things the company might do to support a team member is suggest the attendance of formal training for the team member to increase knowledge or sharpen skills or just provide a better understanding of situation in general. Typically, at least one of these technicians would get chosen to attend a weeklong safety training seminar. These seminars would be held in the winter when outside work was usually slower. The company made it a point to make the seminars not only informative but also fun. Getting picked for participation in this seminar was something that the technicians looked forward to. This experience showed a safety commitment on the part of the company and strengthened the safety background of the attendees.

Example: Accident-Prone Pete

Many years ago, I worked with a chemical manufacturing plant that happened to use a unionized workforce. This particular plant did not have a well-developed safety culture. The accident record in this manufacturing plant needed to improve. Over the course of the year, there were numerous recordable and lost-time accidents, as noted by their OSHA 200 log. Although many workers would typically be involved in work-related accidents throughout the year, one particular worker seemed to be constantly getting hurt.

To provide a bit of history, this particular company had hired four brothers to work at one location. Although the brothers all worked in different departments, they enjoyed status within the union and had the advantage of viewpoints from four different departments. Management was upset with the performance of the youngest brother, who happened to be the most recent hire, because he appeared to be "accident prone." In fact, when I was hired, I could not meet the youngest brother because he was off work due to an occupational injury.

I had heard about the youngest brother because his reputation was well known throughout the plant; however, I quickly met with the other brothers and got to know them on a professional basis. Two of the brothers were members of the fire brigade (which I was responsible for), and another brother held a position of power within the local union. When I asked each brother what his thoughts were regarding the youngest brother, two of them said he was a "good guy," but just accident prone and "young." They went on to say that their brother was not out to take advantage of the workers' compensation system, but that he just got hurt a lot. Two of the three brothers were concerned that their brother would lose his job because of accidents or for some other reason.

Besides talking to his brothers, I talked to the area supervisor and other members of management who had worked with this team member in the past. Members of management agreed that the little brother of the group (let's call him "Pete") was not a bad guy but was lazier than the other brothers and was accident prone (see Figure 8–1).

(continued)

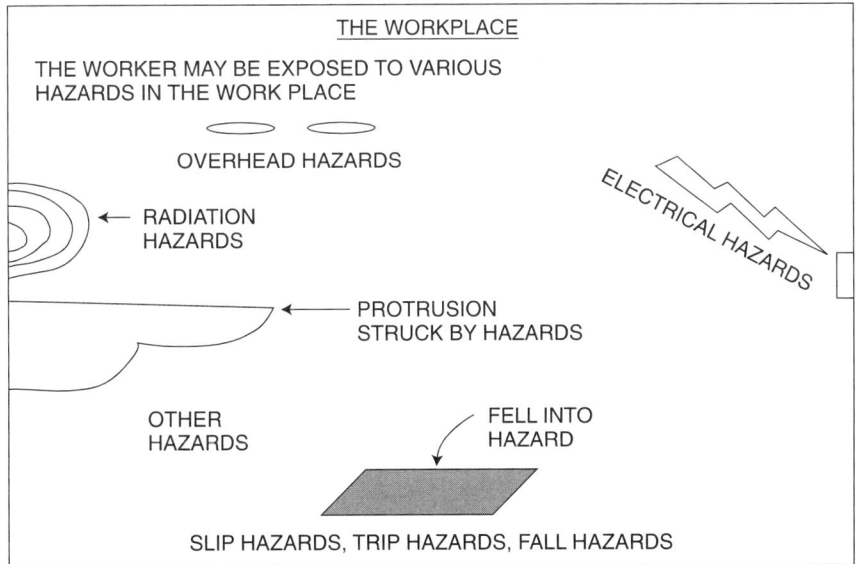

Figure 8–1 Accident Prone. Workers who tend to have accidents while co-workers are accident free are sometimes referred to as "accident prone." Although working conditions are identical for both workers, one will be involved in accidents and the other worker will not.

When Pete returned to work and recovered from his latest accident, I went to his department to meet with him. My impression was similar to his brothers' thoughts about why the little brother was having accidents. Pete seemed to have a good attitude, but he was obviously younger than most workers in the plant and maybe did not carry himself with a lot of self-confidence. In fact, Pete seemed a bit uncoordinated, but appeared to be sincere.

After meeting with Pete for the first time, face-to-face, I proceeded to the human resource department to look through his medical files. In Pete's short career, he had a lengthy medical file. Medical surveillance indicated he was physically capable of handling the work. There were no signs that Pete was involved with drugs, but if this plant had instituted testing for cause, there would have been definitive proof if drugs were involved; however, we will just have to assume that drugs were not involved. (To this day, I truly do not believe that drugs were involved.)

Pete's medical file indicated that Pete was involved in a work-related accident about every two months. Some were recordable, some were first aid, and some were lost time. Although the last accident Pete was involved in was lost time, this was not the most serious accident he had been involved with. So, as far as trends were concerned, it did not appear that the accidents Pete was having were becoming more severe.

I studied each one of Pete's accidents in detail. Several, if not all, of the accidents Pete was involved in seemed to be from lack of focus. On two occasions, Pete walked into two different objects. On two other occasions, Pete fell while either walking on a level surface, tripping over a hazard, or just slipping and falling. Pete's accidents also included cutting himself on several different sharp objects. It seemed that if there were a hazard present, Pete would find it. And if Pete found it, he usually was injured.

I spoke to the human resource manager (HRM) about Pete's safety performance on several occasions. The HRM agreed that Pete appeared to be nonmalicious but accident prone. The HRM was afraid that Pete would either be seriously injured or terminated for cause (unrelated to accident performance). The HRM asked me to consider attempting to place Pete as a union safety representative for Pete's current area/department. The HRM thought that if we could keep Pete's focus on safety, he would be able to steer clear of the hazards. The more I thought about it, the more it truly seemed to be a good idea. Either way, it was worth a try. I told the HRM that I would contemplate the idea and get back with him.

I proceeded to talk to Pete again. I asked Pete why he had been involved in so many accidents in such a short period. I showed Pete a section of the OSHA 200 log and pointed out which accidents he had been involved with. We also talked about the effect that accidents have on workers and companies. We talked about Pete's performance regarding accidents and that other people believed Pete was accident prone. Pete's numbers stood out (in a negative way) from all other workers who had been involved in an accident within the past couple of years.

Pete understood that his accident record was very poor. Pete was also scared that he would be seriously injured or lose his job. Pete was scared, but he did not know what to do to help correct the situation. After all, his older brothers were all able to handle the job, so why shouldn't Pete? He was not about to quit. This was the best job he could ever hope for. He made more money than he ever imagined he would. All of his brothers worked at this plant and could help each other with rides to work. All Pete wanted to do was become successful.

I asked Pete several times why *he* believed he was involved in so many accidents. Pete's response was that "This is a dangerous place. Look at these conditions." I then pointed out to Pete that the working conditions were the same for everyone. I wondered why he, and not others, was involved in so many accidents. Pete's explanation was that he was trying harder than the others to perform his job and that he was focused more than most others in his bid to be successful. I did not have much to say about his bid to be successful, but I did take issue with his focus on safety-related issues and accident occurrences. Pete could come up with no logical reason for his safety performance. Pete believed he was simply accident prone, and he was (along with everybody else) becoming more concerned as time went on.

My dealings with Pete, Pete's supervisors, his brothers, and other members of management became more frequent. On several occasions, I went to discuss Pete's performance off the record with various members of union-elected and appointed officials. The union was also concerned about Pete's performance. To make matters worse, Pete was soon involved in another accident. This one was a laceration from walking into a protruding object. This accident occurrence really concerned me. In fact, it really concerned his brothers and the union. I found out that several union officials had counseled Pete regarding his safety performance and were

(continued)

also disappointed to find out that Pete was again involved in an accident.

At this point, I was ready to give the HRM's idea a chance. After several other conversations with the union, I asked about appointing Pete as a union safety representative. I explained that I believed Pete really was not focusing on safety hazards. My union contact was willing to try the idea, but there were still hurdles to cross.

It seems the union contract only allowed a certain number of safety representatives and a certain amount of time for safety-related business. Some members of management were not interested in expanding the number of union safety representatives. These management members felt that the safety involvement was just another way for workers to get out of doing work. Some members of management were not interested in trying to turn Pete around. He was considered a lost cause, and accident prone. On the other hand, the union was not interested in allowing Pete to take over someone else's occupied spot, even temporarily. They had a full complement of union safety people. They were not about to unseat one of them so that Pete could take over, even temporarily. It seemed as if an impasse had been reached.

Management wanted the union to shift one of their current representatives out of their position so that Pete could do it. The union wanted management to expand the contract and allow an extra union safety representative—Pete. While both sides pursued the discussion, Pete managed to become involved in yet another accident. For a short period, this most recent accident occurrence seemed to drive both sides in opposite directions. Management touted that Pete just was not "mature" enough to handle the plant conditions. The union argued that the conditions were too tough and needed to be upgraded.

Soon a compromise was reached. Management would allow Pete to use some time in representing his department as a safety representative, but only for a temporary period—three months. The contract was not going to be amended, and the union would agree not to attempt to use this temporary situation as a way to increase safety representation.

With the agreement, Pete became a temporary safety representative. I informed Pete of his new duties and was surprised that Pete seemed a bit lukewarm to the idea. In fact, Pete consulted with the union and his brothers before he would agree to be the representative. He soon agreed, however, and proceeded to get input from fellow department workers regarding safety concerns within their department.

Pete went in and around his department and wrote down every potential safety hazard that he could locate. Pete found hazards in places people had not considered before. His thoroughness was amazing. He had amassed five times the amount of safety hazards (or potential safety hazards) that anyone in any department had noticed in the past. He generated a tremendous amount of work orders over hazards that Pete pointed out. Most of the other workers were generally impressed, except for the engineering and maintenance department, whose job it was to address the hazards.

What was even more amazing was that Pete's accident proneness had disappeared. There were no more first-aid cases, recordable, lost time accidents, or even close calls. It appears that Pete's notoriety on the OSHA 200 log was a thing of the past. Pete's career improved, and his propensity for accidents disappeared. Pete's reason for being involved in accidents was a classic case of lack of awareness. Once Pete focused on potentially unsafe conditions, he was able to avoid them. Pete's turnaround in accident occurrences was phenomenal. Pete was not involved in any type of accident from the time of his temporary appointment to departmental safety representative to the time I left my assignment at this company two years later.

Two main things had to happen for this phenomenal turnaround to take place. The first main event was that Pete had to be told in no uncertain terms that accident occurrences are not acceptable to management and that Pete's past safety performance was unacceptable. He needed to be aware that he could not achieve success if he was not successful in safety. Although Pete seemed to understand this point, he could not get turned around without a variety of help. Remember that the counseling that took place informing Pete of his inadequate accident performance was basically an informal performance evaluation.

The second main thing that had to happen was that Pete needed to refocus his efforts and concentrate on the things that were involved in his injuries. In Pete's case, there were hazards present that Pete seemed to find when no one else was injured by them. Something had to be done to open up Pete's eyes and allow Pete to help himself.

Help did come. Pete was given the tools so that he learned how to help himself. With his newfound awareness, he no longer hurt

(continued)

> himself by walking into any objects. It no longer mattered that these objects were not where they were supposed to be. Pete was able to enjoy the team effort. This team included the management, the union, and his family. All of those involved were concerned about Pete's future. As it turned out, Pete's future was a lot brighter than it would have been.

In this example, I was extremely lucky to see a classic case of nonfocus receive the necessary cure. Management is usually able to point out problems, but all too often, the solutions seem to elude us, or the problems just never seem to get solved. All too often, one might come across a worker who just seems to be accident prone.

No managers want this accident-prone worker to be working in their department or on their project because they realize their safety record is in jeopardy. The innovative manager will often find a task for the accident-prone worker to perform at which the likelihood for the worker to get hurt is virtually nonexistent. At the same time, that manager is looking for a way to transfer or get rid of the accident-prone worker as soon as possible. Sometimes termination (voluntary or otherwise) is the final outcome.

Accident Prone

Let me say that accident proneness does not exist. I have used the term in this chapter only because it is a term recognized by most readers. Simply stated, workers who are considered accident prone are those workers who are involved in accidents more often than other workers performing similar activities. Accident-prone workers seem to be involved in these accidents for no particular reason; they just "happen." I cannot accept this statement, and I do not believe the reader should accept this. People who experience accidents because they are "prone" to accidents do not exist. Accidents happen for a reason, not because some people are just prone to them. The reasons must be determined and corrected.

I have long believed that if you properly train and manage workers and provide them with the proper tools to perform their jobs, the workers will be able to perform the most dangerous jobs incident free. In the previous example, Pete had adequate training but lacked focus or awareness of his surroundings. Pete realized that his focus needed to change. Some six months after Pete was incident free, he told me that it helped him to focus on safety when he considered his department a kind of minefield. Pete would take a look around and consider any hazard a land mine. Pete would then map out the locations of these hazards as one might map out land mines. Once this mapping process was complete, he would take appropriate corrective action while realizing that he needed to avoid the area or approach with extra caution, personal protective equipment, or safety systems in place.

The previous example demonstrated that performance evaluations can help build culture in both a union and nonunion atmosphere. Let's go back and try to answer the questions that we asked in the beginning of this section.

Those three questions were:

- Is there a link between safety performance and performance evaluations?
- What part of performance evaluations should be related to safety?
- Should safety performance be linked to performance evaluations?

THE LINK BETWEEN SAFETY PERFORMANCE AND PERFORMANCE EVALUATIONS

A simple answer to this first question is a simple yes. Or at least, I believe that if your company is well managed, the answer *should be* a simple yes. Maybe your company philosophy does not link performance evaluations with safety performance. This is a mistake. If your aim is to further develop or define your safety culture, there is—or at least there should be—a link between safety performance and performance evaluations. So, the answer to the first question is a definite YES. If safety is not part of your general performance evaluations or reviews, you should seriously consider making some changes.

SHOULD SAFETY BE RELATED TO GENERAL PERFORMANCE?

The first question was easy. Let's consider question number two: What part of performance evaluations should be related to safety? Let's try to simply answer *how much* or *what part* of the general performance evaluation should be safety related. Unfortunately, there is no exact percentage that I would recommend, but I would recommend a range of 5 to 20 percent. For your front-line supervisor, performance evaluations should contain 10 to 20 percent safety performance areas. If your general performance forms are 10 pages long, somewhere between one-half and two pages should be safety related.

So, if the company you work for holds safety as a core value, then at the very beginning of your team member's general performance evaluation, there should be at least one safety section. This should directly link safety performance to performance evaluation. The placement of the safety section is important. Even if the length of this section constitutes 5 rather than 20 percent, a message is sent regarding placement of the safety section. The safety sections would typically be part of the initial sections of the performance evaluation.

SHOULD SAFETY BE LINKED TO PERFORMANCE EVALUATIONS?

Let's move on to question number three: Should safety performance be linked to performance evaluations? I believe we answered this question several times over in this section already. If we reexamine the previous two examples, whether we have a union or nonunion workforce, the answer is simply yes.

Myth: Union Workers Should Not Receive Performance Evaluations

Many people believe that members of the union workforce should not receive performance evaluations. After all, increases are negotiated by the union, and individual performance will likely not affect the wages of the union worker. Even though this might be the case, these workers should be given an evaluation anyway. Consider the example with Pete. The first thing that needed to be done was informing Pete that accident occurrences are not acceptable. This was an informal performance evaluation.

Management must let workers know what is important so they can improve. With things improving, success will follow. Continued success usually results in improved benefits and wages for all team members and hence, job security. These things are important to all team members.

So, inform workers regarding their performance. If your particular situation allows you to do it formally twice per year, then do it. If it does not, evaluate all of your workers' performance often, even if it is not documented or formal. Besides performing the evaluation, inform the workers how they can improve.

PERFORMANCE EVALUATION CONTENT

General performance evaluations can vary substantially. These evaluations can be as short as one page and as long as 20 pages or longer. The content of these evaluations can also vary greatly. The general evaluation should touch on all core values of your company. Safety should be one of these core values. The safety section should be the first section, or at least one of the first sections of the performance evaluation.

Whatever your current safety program encompasses could be mentioned and/or discussed during evaluations. Positives such as safety observation programs, attendance, or special safety training or participating in a job safety analysis should be mentioned. Any shortcomings, including notices of safety violations, should also be reviewed during evaluations. Most important, a path forward for safety for the next few weeks or months should be made clear.

Ideally, if your company has been conducting performance evaluations for many years, the reviewer will take out the previous performance evaluations and use them as a guide. Although these evaluations encompass many areas, we will only be discussing safety areas. The previous performance evaluation should have listed both strengths and weaknesses. For weaknesses, a path forward to assist the worker in strengthening a weak area should be outlined. Such suggestions might include the following:

- Participation in a mentoring program
- Attendance at after-work organizations such as Toastmasters
- Completion of in-house seminars or training
- Completion of adult education programs
- Completion of college or other classes
- Other appropriate activities

If your current performance evaluation program does not include any of these items, it is time to consider changes. Start with your next performance evaluation. You may not be able to go to the last performance evaluation and determine if the team member is still following the path forward that was outlined during the last review. The reviewer, in a joint effort with the team member, should determine the path forward. Once the path is chosen, the team member is expected to travel the path with only minimal guidance. During the next formal performance evaluation (which should be no longer than six months), the team member and reviewer will review what has been accomplished.

The phrase *out of alignment* is used when the specified path has been strayed from. The reviewer and team member would jointly determine if the path that the team member is currently traveling is appropriate. A certain amount of twists and turns will likely be present in any road, especially if the road is a long one. But if the general direction is veering too far from the goal, it might be time for realignment. During the performance evaluation, the reviewer and the team member will:

- Determine where they are
- Determine where they want to go
- Choose a path to get from where they are to where they want to be
- Travel the path in the right direction until the goal is met

For each subsequent performance review, the same basic points are covered. If the team member strays too far from the path, realignment may need to take place. If goals were set, were those goals reached? If they were reached, should new goals be assigned? If goals were not reached, can we determine why? Should goals be reestablished or new ones be set? The basic principle behind performance evaluations seems to revolve around the Hawthorne Effect that we discussed in preceding chapters.

We evaluate our team members' performance (on safety and other areas) because our people are important. We need all of our people to be in alignment and to work toward a common goal. Team members need to know what path they are expected to take. They also need to know when they are straying from the chosen path. If they need to be aligned, they should be advised of this need and assisted with the realignment effort. When used appropriately, performance evaluations are a valuable management tool. This tool is too important for any team members to be able to say: "I can't remember the last time I had a performance evaluation. It must be at least five years." Or a statement like this: "I don't get any type of evaluation, but sometimes my boss calls me on the day my raise shows up on my paycheck and tells me what percentage increase I received this year."

If people are our most valued asset, then these two statements should never be made. Unfortunately, all of us have heard statements similar to those on numerous occasions. The grim reality is that all managers do not practice the belief that team members are our most valued assets. If they did, it would be difficult to find team members who could make statements about never or rarely having performance evaluations.

SUMMARY

Performance evaluations are a tremendously powerful management tool. Whatever policy your company has regarding performance evaluations should be communicated with team members up front and adhered to. Everyone, no matter what position they hold, needs to know where they stand and how to reach the next level in their careers. Besides reaching the next level team members need to know how they are doing now, and what management believes regarding past performance.

It is important to note that the performance evaluation and the salary review are different and are meant to be different. A salary review might take place once per year, but performance reviews should take place often, dependent on the particular circumstances. The performance review is an assessment. If a team member's performance is sub-par, yet no one advises that team member of this sub-par performance this sub-par performance will likely continue. Allowance of sub-par performance is a form of negative reinforcement. This cycle must be turned around to positive reinforcement. A path must be made toward continuous performance improvement. This path includes informing the team member of sub-par performance and working together with the team member on a path toward performance improvement.

There are those who believe that performance evaluations have no place in the union environment. I do not agree. I believe that some type of performance evaluation, whether highly formalized or informal, is important whether unions are involved or not.

9

Safety Incentive Bonus Programs

Safety incentive bonus plans have become part of the safety culture for many types of businesses. They are commonplace for those types of businesses that have a large percentage of their workforce in the field, shop, or on projects that involve physical work. Safety incentives are less common in businesses that have a large percentage of their workforce in office-type jobs. Like any incentive, the basic principle is to share in success. For baseball players, they might have an incentive that revolves around their batting average. For instance, the baseball player might receive a certain amount of extra money if his batting average increased 10 percent. The team might agree to such a clause in the contract, believing that if that particular player would raise his batting average 10 percent, the team would probably be more successful.

Lots of people participate in incentive programs. If your current company has a profit-sharing program, this is a type of incentive. The idea behind profit sharing is that if every team member works hard and the firm reaches the goals as stated in the profit-sharing plan, then everyone would share in the good fortune and see the value of the profit-sharing plan increase. Sales and marketing departments are known for individual and group incentive plans. If the salesperson sells jobs over a certain dollar amount, that salesperson would get a reward. Or if the department would reach a certain goal, the department would reap the benefit.

For safety incentive bonus programs, I believe that most of the focus should be on the team rather than the individual. Although every team member has his or her specific role in the big picture, I like to see incentive plans that are based on group performance rather than individual performance. If we remember the Hawthorne Effect, this philosophy fits in with it. That is not to say that some individual recognition is not also in order. We will talk about safety recognition on an individual basis at the end of this chapter.

Let's take a closer look at how these programs are set up or should be set up. In a well-designed incentive program, we should see that each participant shares in the rewards of good safety performance. Let's look at who should be eligible and to what extent team members should reap these rewards for good safety performance. For the benefit of this discussion, these programs are described in this chapter as "lucrative" bonus programs.

A particular lucrative program with which I was associated consisted of a cash bonus for almost all of the department members. The workers who actually performed the work were eligible to receive one day's extra pay per month. Key operators and lead personnel (both of these positions were union and not considered management) would receive up to two days pay per month depending on the general foreperson's recommendation. Supervisors would receive one week's pay per month if all of their reports qualified. General forepersons and above were not eligible for a safety bonus.

This program incorporated some excellent principles yet still seemed to be a bit lacking. One of the strong points was that this program allowed many of the team members to participate. It stressed the importance of team members being safe. It rewarded those workers who were not involved in lost time accidents during the month. It stressed to the supervisor that if his or her direct reports were accident free, then there would be a substantial reward for the front-line supervisor. Although certain management principles downplay front-line supervisors and their roles, in this company, management felt that the front-line supervisor was the key to success—not only in the safety arena, but in most other aspects of the work.

One important point not mentioned in the previous discussion was that this was not just a safety bonus, but also an attendance bonus. The attendance part was that workers must show up on time for work every day. They were allowed to be tardy twice and still qualify, but an unauthorized absence disqualified both the worker and the supervisor. Also, for the operator and a lead person, the recommendation for the bonus seemed to be more political in nature rather than based on objective data. The following are some of the principles that this particular lucrative safety bonus program was based on:

- Field supervisors (front-line supervisors) are the most important link in the safety chain.

- Field supervisors should have an incentive based on performance goals of their crews.
- Hourly worker incentives should be based on individual performance and not on the team.
- The upper ends of management should not participate in a safety bonus but instead should concentrate on profit and can participate in a profit bonus program.

I cannot totally agree with all of these items. Are field supervisors the most important link in the safety chain, or is it the team members? I believe that although a strong case can be made for the most important link in the safety chain being the front-line supervisor, one might argue that, in reality, the top management person sets the tone for safety for the whole company. Therefore, the top manager should be considered the most important link in the safety chain.

A CHAIN IS AS STRONG AS ITS WEAKEST LINK

The chain is only as strong as its weakest link, and the weakest link in any organization can vary. The workforce at any particular place of business is different. Your company may have the best-equipped and most dedicated hourly assembly-line team member workforce in the world. Possibly the company you work for may boast of the skill and ability of your front-line supervisors and middle management. Or your company may have the most highly touted chief executive that every Fortune 500 company is begging to lure away. I believe that it is truly difficult to say what group is the most important link in the safety chain, but it depends (see Figure 9–1).

Although not always the case, I believe that having a strong supervisory workforce is an area where most organizations are weak. Again, I am not saying that the field supervisor's role is the most important or the least important, just that many organizations would be uplifted by bolstering their supervisors through grooming and training. Just like everyone's role on the team, the role of the supervisor is important. I believe so strongly in the need to train and groom front-line supervisors that I dedicated an entire chapter to this important subject.

So, the answer is that weakness could be anywhere. Any area of weakness needs to be determined and shored up. The subject of determining weakness and implementing a plan to shore up the weakness is discussed

Safety Incentive Bonus Programs   **129**

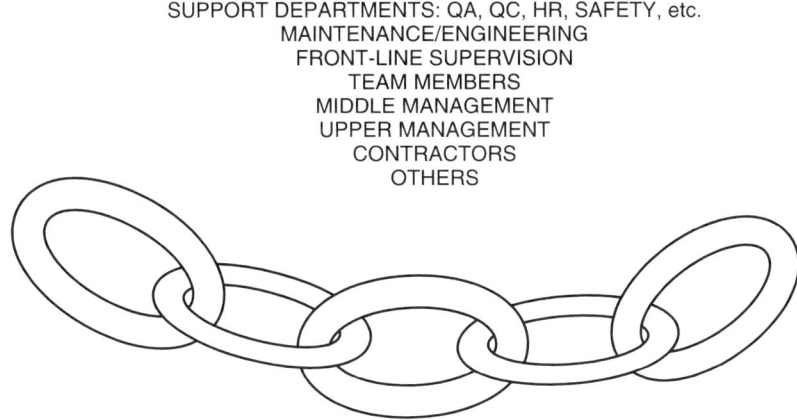

Figure 9–1 The Management Structure Chain. A chain is only as strong as its weakest link. Depending on the company culture, that weak link could be almost anywhere.

later on, but first let's discuss what the roles of the supervisor, team member, and top management should be.

Front-line supervisors should set the tone for safety, lead by example, ensure that workers are properly trained and qualified to perform their work, and provide the proper tools, personal protective equipment (PPE), planning, and so on to ensure that the work is done safely and properly. Front-line supervisors have a difficult and important job. If we relate back to our discussion of the Hawthorne Effect, supervisors need to be treated as important team members also. They interface between the team members who are getting the work done and the management team. This situation can be difficult. Front-line supervisors typically wear many hats. They must manage the job from a viewpoint of the superintendent, yet they must deal with the workers on a personal level on a constant basis. The supervisor might be expected to be part of all of the following job functions:

- Human resource manager
- Safety director
- QA/QC manager
- Nurse
- Industrial engineer
- Maintenance person
- Computer programmer
- And a variety of other functions

The hourly worker has an important function. Hourly workers are important team members whose skills and abilities are a substantial asset to any company's success. Everyone needs to realize that he or she is an important team member. If we look at statistics, hourly workers are the most likely group to be involved in work-related accidents. If everyone shares the modern philosophy regarding accidents, then we all believe that all accidents are preventable. We would also believe that the lack of accidents coincides with effective management. This effective management must take an active role in ensuring that team members (hourly workers) understand the tasks they will be expected to perform and how safe performance of each task relates to their individual job responsibilities.

The team member must take on working safely as part of his or her responsibility. Organizational management must ensure that they communicate that working safely is a condition of employment. Working safely must be a core value that is known and shared by everyone.

Hourly workers must follow the lead of their front-line supervisor and take on the responsibility of working safely on those occasions when they are not closely supervised. For most of the working day, a supervisor will not be on the scene playing traffic cop. The modern front-line supervisor should not be looked on as a traffic cop with a radar gun, sitting in a speed trap, waiting to give out speeding tickets. In the modern work relationship, the supervisor is more of a coach working with the other coaches and team members for the success of the organization.

The manager is like the head coach. The manager works with the coaches (or supervisors) from the various departments to make sure that the team is playing in unison. If the team is not operating like a well-oiled machine, then the manager will make adjustments or take corrective action. The manager leaves the day-to-day interaction of team members to the specialty coaches but may interact with a team member on rare occasions. The manager understands the importance of safety and ensures that the coaches and team members get the message and act accordingly.

So far, we have discussed the role of the supervisor, hourly worker, and manager within one of the more lucrative safety incentive programs that I have been involved in. Then we looked at what the role of the supervisor, hourly worker, and manager should be. Now, let's discuss what safety incentive plans should look like for both management and hourly team members.

As previously discussed, management's role in safety is different than the hourly team member's role. This basic difference sometimes does not lend itself toward having all groups participating in one incentive plan. Therefore, the incentive plans for management and hourly team members must be different. Some of these differences are discussed in more detail as follows.

Management's Incentive

Management's incentive should be based on safety performance goals for the division or the company as a whole. This point is important. Management needs to keep the "big picture" viewpoint. If we look at a business and its stock price, the stock price will typically rise when the business is performing well. In order for a business to do well, all of the pieces that go into the end of the product need to be coordinated by management.

If we are making widgets and our widget is no longer being produced better, safer, cheaper, and faster than our competition, then we will likely lose market share and be less profitable. Management needs to determine any reasons for product failure and adjust accordingly. If we draw the analogy of a football team, you will have better success when your offense, defense, and special teams, front office, and coaching staff are all working toward winning a Superbowl.

Hourly Incentive

Hourly safety incentives should be based on both individual and team performance. The hourly team member will have a different role in the organization's success. This person will perform little coordination among departments but will concentrate on individual efforts and hands-on activities. Analogous to a football team, this person is on the field as a position player. He might never handle the ball, but nonetheless he has an important part in the team's success. This person will focus on individual performance and on the team performance.

Let's go back and examine the lucrative safety plans at the beginning of this chapter. Let's discuss what could be done to improve these programs. We will look at management's role first. Notice that top management was not included in the lucrative safety incentive plan—only front-line supervisors, lead persons, and operators. Also notice that upper-level management personnel were not allowed involvement with the safety incentive program, but were involved with the profit bonus program. This sends a mixed signal to upper management, front-line supervisors and lead persons, and the whole team.

How can upper management participate in the profit bonus program but not the safety incentive program? To make the situation worse, only the upper-level managers were eligible to participate in the profit bonus program. Could it be that upper management cannot be bothered with what the lower-level people are concerned with? Probably not, but I think that some people might see a correlation. If your team members see the correlation, you can understand that this perspective is not good for a teaming atmosphere.

Not having upper management involved with safety incentives is probably not a good way to promote teaming. Remember the Hawthorne Effect: If we can show our people that they are important team members, it will lead to a more productive and successful company. If we do not treat our people like important team members, this will not improve our current situation. Having two different incentive plans that two different groups can participate in will probably drive these groups apart rather than promote teaming.

Notice that front-line supervisors were awarded one extra week's pay per month if their crews met the criteria. If the front-line supervisor worked a 40-hour week, this would equate to almost a 25 percent bonus per month. I think a bonus of this magnitude would be a great benefit for anyone who would be eligible to receive it; however, what I saw in the field is that front-line supervisors would bend the rules to ensure that they received the bonus. If they had a worker who was going to disqualify them from receiving their bonus due to tardiness, it would be human nature to have this person terminated or transferred to another department or to use many other methods rather than be disqualified from receiving the bonus.

These lucrative incentive plans, although well intentioned, really need to be restructured keeping the following information in mind: The plan, as discussed earlier in the chapter, has managers not able to participate in the program as lower-cchelon supervisors, lead personnel, and the like. This needs to be adjusted. Supervisors, operators, workers, leadpersons, and managers should be in a plan that promotes teaming. When different levels of awards are given to different workers in different job classifications, you are deviating from the teaming concept. The teaming concept is a key to increased productivity and success.

Another adjustment that needs to be made is that the incentive plan covered not only safety but also absenteeism and tardiness. I agree that all of these things are important. How can a business be run if everybody shows up late or nobody shows up at all? Obviously, no business could run without people to do the work. But safety incentive plans should deal just in safety. They should not try to couple the plan to include tardiness, quality, or other aspects of the business. If our premises are correct, as our safety culture improves, so will business in general. Running an effective safety program and building a safety culture will reflect as a general improvement in management.

CASH AWARDS

Cash awards should be avoided, especially when large amounts of cash are being used for awards. Don't misunderstand me, cash has its place. For certain members of the team, it may be the only acceptable incentive. But for safety awards programs, cash is not the best incentive. So, the lucrative incentive plans described in the beginning of this chapter need to be adjusted. We need to remove the lucrative cash awards and replace them with more reasonable awards such as articles of clothing.

Safety celebrations for reaching group or individual safety milestones were not included in the incentive program mentioned earlier. Safety celebrations, although not considered by all to be part of an incentive plan, certainly represent a cash outlay from the company that benefits the worker. So, in a certain manner of speaking, safety celebrations are a type of incentive.

GROUP SAFETY MILESTONES

Group milestones need to consider the big picture and the microcosm. In other words, we need to analyze the whole company or division and each small group within the organization. You will likely find that certain departments or job classifications jump out as experiencing a much higher incidence of accidents than others.

Example: The Job No One Wants

> Certain 100-pound bags of chemicals were purchased for sale and distribution. These bags came in a boxcar containing 100,000 pounds (1,000 bags) per railcar. Instead of being "palletized" (20 or so bags
>
> *(continued)*

prestacked per pallet), they were typically "loose" (stacked directly on the floor with no pallet below them). It was soon determined that manually placing these bags from the floor of the boxcar onto a pallet was considered a simply terrible job. The purchasing department advised that this was the only way these bags could be purchased.

The warehouse personnel found out that competitors were buying this material already palletized and decided to approach purchasing again. When the warehouse personnel approached the purchasing people with the newly discovered information from a competitor, the response appeared to change. Purchasing then insisted that the only way the company could realize a profit on this material was to buy the material in 100-pound bags on the floor of a boxcar. A credibility gap now existed. The warehouse personnel were not satisfied with this explanation, and trust became a factor. The job of palletizing these loads met with stiff opposition. Warehouse personnel would typically "call off" on days when this job was assigned. This practice put a strain on an already thinly staffed department.

Management tried to lessen the effect of this job by creating a fair job rotation. Everyone tasked to do this job was assigned to it for a two-hour stint. There was still dissatisfaction, however, because some workers were very productive during their two-hour stint, whereas others were much less productive. The plan was changed to a quota or piecework system. Each worker was assigned a task to palletize four pallets of bags. This system did not last long either. It turned out that some workers would finish their tasks quickly, but some could not finish by the end of their shift.

Management decided to start an incentive plan just for this task within this department. The powers that be thought that this would be a positive way to get this task done better, safer, cheaper, and faster. A meeting was held to kick off the new program. Management approached the group with all the positivity they could muster. The warehouse personnel politely accepted the program because it provided them with an added benefit but informed management of their concerns. People from other departments also voiced their dissatisfaction. They did not think it was fair that certain warehouse personnel would get this incentive while they were not eligible.

Lessons Learned: A Team Dissolved

Management quickly abandoned this incentive plan. The problem seemed to disappear when the bags were ordered palletized rather than on the floor of the boxcar. Management realized that the departmental incentive was not the best alternative. Instead of having the desired positive effect, it created dissension in several departments. If management believes in the Hawthorne Effect, then teaming needs to be continually promoted. This incentive promoted teaming on a small scale but actually pitted different departments against each other.

The warehouse personnel made it clear that the bags should be ordered palletized. Unfortunately, their requests were not acknowledged. Mistrust between purchasing and the warehouse personnel escalated when it was discovered that, in the minds of the warehouse personnel, purchasing had initially lied about palletization. If we were trying to demonstrate that the warehouse personnel were valuable team members, this experience with the bags sent mixed signals. How much would one workers' compensation claim for an injured back have cost this company? I am not going to answer that question in dollars and cents, but the cost in safety culture development was excessive. The bitter feelings over the handling of this event were discussed for years.

Example: The Misplaced Celebration

Management learned that assembly personnel in the widget assembly department, step number 6, experience five times the amount of accidents as in the other steps. Management decides to implement a special milestone celebration for members of this department. Their hopes were that once people realized that milestones were reachable and that management recognized members of the department for reaching these goals, the safety performance would continually improve.

Although this situation is hypothetical, it is similar to the 100-pound bag example. This situation should be avoided for the same reasons that the 100-pound bag problem demonstrated. There is usually a straightforward reason why accident incidence is high in a

(continued)

certain department. I do not believe that using an incentive program or safety celebration to solve poor safety performance is the primary way to remedy this type of situation. We will discuss different tools to use to analyze poor safety performance and ways to improve poor safety performance in later chapters. Safety celebrations have their place; they may not be a part of an incentive program, but they probably fit into some part of the program.

An effective way to use group milestones is with safety celebrations; however, the safety celebrations should recognize the entire group and look at the big picture. If you have a business that has various remote locations, the safety celebration should include not only those team members who are stationed from that location but also selected persons from the home office who were instrumental in the project's success. Selected persons would include all of those team members who were closely associated with the project. Someone needs to make a decision regarding who gets invited and who does not. My philosophy is that if there is doubt about whether team members had enough involvement to be invited, invite them. If those team members located at a remote site were truly members of the team, or even if they think they were members of the team, invite them. (Safety celebrations are also discussed in the last chapter of this book.)

Recognition and awareness are the two keywords for incentive programs. The size or amount of the award is secondary or tertiary when compared with recognition and potential gains in awareness. When you take a closer look at recognition and awareness, they are ways to further teaming and show team members they are valuable.

INDIVIDUAL INCENTIVE AWARDS

Let's shift gears a bit and discuss individual incentive awards. I am referring to the gift given to workers who celebrate a safety milestone.

Many years ago, I worked for a chemical plant that gave workers gifts for completing 1, 3, 5, 10, 20, and 25 years without a lost time accident. These gifts were similar to what was given to team members for successful years of service. The awards for safety were similar to the awards for service—the longer the milestone, the nicer the gift. When it came time for a safety award to be distributed, it was typically done at a company function such as a staff meeting or safety celebration. Although these awards were for individual performance, teaming was accentuated and awareness and recognition were always emphasized. I believe that these individual awards, when administered properly, can have a positive effect on safety culture development.

SUMMARY

Safety incentive award programs have become commonplace in the industry in recent times. There are two types of incentive awards: the award that is based on the group's performance and the award that is based on an individual's performance. Both of these types of awards have their place and can be effective when used properly; however, attempting to use an incentive award as the sole means of accident prevention will probably result in a less than successful endeavor. Incentive plans should be used as a means to increase awareness and recognition. They should not resemble the jackpot for the local lottery. Cash prizes should be discouraged. Teaming and the Hawthorne Effect should be kept in mind when setting up incentive programs or adjusting current programs. An effective program emphasizes teaming and involvement at many levels.

10
Planning for Field Compliance

Success in compliance is not reached through a haphazard approach or because of dumb luck or chance. Attaining field compliance is the result of hard work and follow-through by a dedicated team. In this chapter, planning for field compliance is the main emphasis. Success in reaching field compliance starts with a plan. Even though having a good plan will not guarantee field compliance, it can put the safety program on the right track and will increase the chances for success.

Some field compliance issues derive their basis from guidelines promulgated by outside regulatory agencies, such as the Occupational Safety and Health Administration (OSHA), the Environmental Protection Agency (EPA), the local fire department, or a local health department; however, internal issues related to environmental, safety, and health concerns, such as ergonomic assessments, that a company adopts are often not specifically covered by outside agencies. On the federal level, OSHA does not yet have a specific ergonomics standard. Many companies that are concerned about cumulative trauma disorders (CTDs) put together internal standards to address these issues. These internal company standards can include items such as:

- Workstation assessments for ergonomic concerns
- Job safety analyses (JSAs) specifically studying CTDs
- Committees to approve purchases of workstation furniture, tools, and supplies for ergonomic concerns
- Attendance at ergonomic awareness or other related training

Ergonomics is just one area in which many companies have adopted stringent internal requirements. Companies concerned about indoor air quality might have similar inside compliance requirements. Again, indoor air quality as it relates to "sick building syndrome" is not specifically regulated as part of a federal OSHA standard. Although guidance and recommendations for allowable limits for some chemicals in the air are available, no specific regulation addresses air contamination at the levels found to create the usual complaints within office settings.

Besides areas that are not specifically covered by OSHA, some companies adopt internal standards that are more stringent than OSHA or regulations. Take a company whose workers perform tuck pointing. The OSHA fall protection rule takes effect at six feet above ground level. Some companies have adopted a more stringent internal rule such as a fall protection requirement at five feet above ground level. Many window-washing companies that work on highrise buildings provide extra safety procedures not required by OSHA to ensure worker safety.

Inside compliance issues can originate from a variety of different departments. The different departments that may have safety-related compliance issues could include the following:

- Safety
- Quality assurance
- Human resources
- Loss prevention
- Medical
- Security
- Legal
- Environmental

The term *field* might give one the impression that I am referring to the great outdoors—a place where cornfields have been converted to some other use. But for the purposes of this discussion, the *field* is anywhere that work (not farming work) goes on, including many office spaces. Let me offer some examples.

One of my steadier customers used to be a large financial services company. Most of the workforce at this financial services company included office-type workers who performed most of their hours at work on the phone or in front of a computer terminal or both. They have a huge customer service contingent, along with thousands of people who work in the collection department. Besides the office workers, this company has a small maintenance staff that takes care of physical facilities. Services to this company include heating/cooling maintenance, clean air, janitorial services, and small repairs. For this company, the field is actually office space. Their safety issues included:

- Indoor air quality
- Fire extinguishers
- Storage near sprinkler heads
- Fire drills
- General hygiene
- Hazard communication

So the field is not always the field per se. For a financial services firm, the field is actually what most other people would consider the office. There are safety-related issues in the field no matter what type of business one is in.

The field is typically anywhere the company does its work and experiences its most serious safety-related concerns. For a manufacturing facility or factory, the field is the shop floor where the work is going on. For a warehousing or storage operation, the field is the warehouse and loading dock. For a construction company, it is most likely the active construction sites. For a pharmaceutical company, it might be the laboratory, packaging, warehouse, shipping, and receiving departments or a variety of other places. For every business, the exact location of the field will be different, but even though the location of the field might be different, the principles guiding compliance are the same.

After determining the boundaries of the field and the type of work activity going on within those boundaries, one should be able to begin determining what the field compliance issues may be. If the facility we are touring appears clean and well organized with no obvious safety hazards, then the odds are good that this facility has a written safety program.

In a well-developed safety culture, there will typically be two types of documents used for planning field compliance: a safety program (or general safety manual) and a site-specific safety plan. These two documents are sometimes intertwined and used interchangeably, even among safety experts, but these two documents are very different. These two key planning documents will aid in attaining safe work performance.

Planning for field compliance takes two different types of planning documents. There is typically only one safety program document, but many site-specific safety plans. Or sometimes an organization will incorporate a variety of departmental rules or site-specific guidelines within the general safety program document. No matter how this is accomplished, the two different levels of planning documents will exist. Let's first discuss the safety program and then the site-specific health and safety plan.

THE SAFETY PROGRAM

The safety program provides written guidelines to ensure that work is being performed up to the company standard. These written guidelines may consist of various documents from various departments that are general procedures indicating how work will be performed safely. These written general guidelines when taken together constitute what is referred to by OSHA as a safety program. The safety program should:

- Document the company safety philosophy.
- Outline the general role and responsibility toward safety.
- Provide general guidance on how work will be safely performed.

Remember, the safety program is a *general* document that spells out the *general* safety principles by which a company operates. Some companies have a document called the safety manual. Depending on the contents of the safety manual, it may fit OSHA's definition for a safety program. Safety programs usually change less frequently than site-specific safety plans. We will see some examples of how these two types of documents interrelate later in this section.

In a safety manual or safety program, one should be able to find general guidance for the safe operation of equipment and the conduct of all work activity. Although these manuals can be set up in a variety of ways, I prefer to have these manuals begin with a "mission statement." For a mission statement to be effective, it may contain a signed, two- or three-paragraph statement from the owner or chief executive documenting that person's and the company's overall commitment to safety.

The contents of the safety program can be extensive. If your business deals with highly hazardous equipment, conditions, or work activity, the safety program could be many volumes. In fact, for some large government-run hazardous and mixed-waste (radioactive and hazardous) sites, the safety program *is* several volumes. For these same sites, the site-specific safety plan is also extensive. These plans are typically many-tiered levels of many volumes. If your company engages in the more complicated and hazardous work activity, it will likely have a larger and more complicated safety program.

If your company is a metal stamper or manufacturer, you will probably have a lot of concern over machine guarding issues. In fact, you might have a volume or more regarding punch press guarding, shears, assembly lines, paints and finishes, and more. If your company deals more with construction, you will likely have a concern regarding steel erection, tying off, trenching and excavation, and other issues. If your company does office or computer work, your concerns probably include ergonomic and office safety issues.

The point is that every company's safety program should be different because every CEO's commitment to safety is different and every company's culture is different. The important principle to remember is that the safety program for your company should address general safety hazards that your team members will encounter. The more hazardous the work, the more extensive the safety program will be.

The safety program will outline *general* guidelines, not specifics. The specifics will be addressed in the site-specific safety plan. The safety program will state general culture items such as: "All work will be done in compliance with current OSHA guidelines." If your company is progressive, it will go above and beyond OSHA guidelines and institute more stringent guidelines.

Example: Exposure Limits

> When dealing with issues of flammability, let's say that OSHA uses a limit of 10 percent LEL (lower explosive limit) as a safe level. In this case, a more progressive company may believe that the OSHA level will not provide employees (or the company) with adequate protection and chooses to adopt a level of 5 percent LEL or lower.
>
> As another example, let's use chemical exposure limits. OSHA uses a PEL (permissible exposure level) of 1 ppm (part per million). A company might find it advantageous to choose a level of less than 1 ppm to ensure that exposure levels stay low. Taking a stance that is more stringent than current government standard levels provides companies (and workers) with an extra level of confidence that workers will not be injured or have their health adversely affected by exposure.
>
> Some companies believe that setting their sights on levels that are more stringent than government guidelines ensures that even if a safety system fails, workers will not be overexposed and the company will not be placed in a compromised situation.

Some in-house guidelines are not based on government regulations but on company in-house expertise or general experience. Over time, successful companies that repeatedly perform work activities develop expertise. With this developed expertise comes the development of in-house guidelines. These in-house guidelines become documented, formalized, and finally become compliance issues.

OSHA has published a DRAFT rule that deals with safety programs. Under this proposed rule, each employer must "set up a safety and health program to manage workplace safety and health to reduce injuries, illnesses, and fatalities by systematically achieving compliance with OSHA standards and the General Duty Clause." The proposed rule is called the **DRAFT PROPOSED SAFETY AND HEALTH PROGRAM RULE 29 CFR 1900.1 Docket No. S&H-0027**. There are five core elements that OSHA proposes the programs have:

- Management leadership and employee participation
- Hazard identification and assessment
- Hazard prevention and control
- Information and training
- Evaluation of program effectiveness

These five points are in basic agreement with the theme of this book; however, the current standard appears to be somewhat vague. For example, the following quote is taken directly from the proposed rule: "The program must be appropriate to conditions in the workplace, such as the hazards to which employees are exposed and the number of employees there." I am hopeful that before this standard becomes law, it will contain substantially more definition and detail.

I have studied this proposed rule and was disappointed when I observed that nowhere in the proposed rule are the terms *JSA* and *site-specific safety plan* mentioned. When you look at the description of how the employer is expected to maintain a safe and healthful work environment, however, both the site-specific safety plan and the JSA seem to be alluded to. The main body of this proposed rule is included in Appendix C of this book for your review. As previously stated, the OSHA guide to JSAs has been included as Appendix A of this book.

SITE-SPECIFIC SAFETY PLAN

Besides the general safety program, a site-specific safety plan is needed that provides the necessary details to ensure that the principles in the program are put to use. The plan should state the names of the individuals who are responsible to implement the work safely. The plan should have certain key components, including (but not limited to) the following:

- Details on hazards/control measures
- Air-monitoring requirements
- Site-specific emergency information
- Material Safety Data (MSD) sheets or other site-specific pertinent information
- Job safety analyses
- Training and/or orientation information
- Other pertinent information

Now that we have discussed both the *general* safety program and the *site-specific* safety plan, let's look at some examples.

Let's consider the relationship between the safety program and the site-specific safety plan to confined-space entry. For a typical company that has a general concern for confined spaces, this should be mentioned in the company safety program. Besides being part of the safety program, if confined-space entry is a concern at a specific work location, it should be mentioned in the site-specific safety plan. Confined spaces are mentioned in both the safety program and the site-specific safety plan. The safety program contains a general procedure outlining how confined spaces will be addressed. The typical information will include the following:

- The definition of a confined space
- A flowchart and other information to aid in determining that a space is a confined space
- Information to determine whether the confined space is permit required
- Training requirements for all of those personnel associated with the confined-space entry
- General hazard determination
- Minimum air-monitoring requirements
- General approval requirements
- General first-aid/CPR requirements
- General requirements for rescue
- A blank confined-space entry permit
- Other pertinent general safety information

The site-specific safety plan will likely contain a copy of the general confined-space procedure from the safety program, along with the following:

Planning for Field Compliance   **145**

- The specific location of the confined space
- A specific description of the confined space, including means of egress, contents or previous contents, lockout/tagout considerations, etc.
- The specific team members who will be tasked to enter, monitor, supervise, or otherwise be involved with the entry
- Current training certificates for all those involved in the entry
- The specific team members who will be qualified to perform first aid/CPR for the entry
- The information used to determine that the space fit the definition of a confined space
- Specific rescue information, such as who will perform rescue
- Current certifications or qualifications of the rescue team
- Other pertinent site-specific required information

This other pertinent site-specific required information could be a certain section of the site-specific plan or could likely be in the form of a JSA. JSAs were discussed in detail in Chapter 7, "Training." JSAs were also mentioned in the initial chapters of this book and will continue to be referred to at various times throughout many of the remaining chapters. The JSA was mentioned as a training aid. It is mentioned here as an important part of the site-specific safety plan. As safety culture develops, each site-specific safety plan will contain more and more JSAs, or the JSAs that are included will contain more useful information as they evolve. Depending on the development of your safety culture, you might include a JSA covering the confined-space entry itself. Besides this JSA, you might include the JSA covering the purpose of the entry.

If the entry into, for example, a storage tank is to inspect and replace mixer blades that are in the middle of the tank, then you might find a variety of JSAs, or you might find one larger JSA that continues to evolve. If you have multiple JSAs, the first JSA would cover confined-space entry. Another one might be for cleaning out the sludge or leftover material from the equipment so that it can be properly inspected. Yet another JSA would be included for inspection. If the inspection reveals that other work such as replacement of blades or equipment is needed, another JSA may be required. Instead of creating separate JSAs, this could be done with one basic JSA with amendments or separate sections with specific work task breakdowns. Depending on the size and type of organization, you might have three different crews performing these activities.

The first crew might specialize in entries and provide the "hole watch" and monitoring necessary for testing and entry. The next crew might consist of outside cleaning contractors who enter the tank for the express purpose of shoveling out the sludge and/or cleaning the interior of the tank. The next group might not enter until after the tank has been fully cleaned. This might be some engineers or maintenance people who have been tasked to inspect and/or make repairs to the tank.

If we look at the different types of hazards involved in the entry, cleaning, and inspection and repair, we realize that the work tasks include different skills, and the hazards that team members might encounter while performing each of these different work tasks vary substantially. For people who work for the smaller company, having this much specialization might seem wasteful, but for people who have worked in midsize to large chemical companies who routinely perform this type of activity, a setup with at least three crews is commonplace. These three crews would include an entry crew, a labor crew to perform cleaning, and the maintenance/engineering crew.

The entry crew might include team members familiar with the operation of air-monitoring equipment and chemical contamination. The labor crew who enters and actually performs the cleaning activity can often consist of outside contractors who specialize in chemical cleanup. This part of the job is typically labor intensive, hot, dirty, and contains a lot of back-breaking, heavy manual labor work. Although I have seen a certain amount of this kind of work performed in-house, I have observed that most of the dirtiest, smelliest, and sometimes most dangerous activities are performed by subcontractors. Once the cleaning has been completed, the engineers and maintenance people will enter and perform the inspection and replacement activities.

Besides the three JSAs connected to the three distinct tasks, more JSAs could be necessary. If the inspection requires welding to go on inside the tank, this should be included in a JSA. This could have a tremendous effect on the work tasks, the hazards, and the JSAs.

Another hazard analysis that should be done in writing is what should happen during an emergency. This emergency response planning might be outlined in the JSA or might be part of the site-specific safety plan. The questions of what should happen during an emergency should always be asked and answered beforehand. The JSA or site-specific safety plan would outline the procedures to be taken if an emergency occurs. In a well-developed safety culture, rescue from the tank would be spelled out. This would include the who, what, where, and how of rescue. This JSA or written hazard analysis would include the following components:

- Whether the rescue would be horizontal or vertical
- Any requirements for entrants to wear rescue lanyards, etc.
- Other pertinent rescue-related issues

Clearly, JSAs are an integral part of our safety program and need to be included with the site-specific safety plan. The more developed the safety culture in our organization is, the more likely that many JSAs will be written to describe a certain work task. Let me emphasize that the JSAs are not part of the safety program. The JSAs are part of the site-specific safety plan. Let's look at some more examples of how the safety program and site-specific safety plan are interrelated.

Let's look at hazard communication (HAZCOM) from the standpoint of the safety program versus the site-specific safety plan. For HAZCOM, companies will likely have the key components specified in the safety program. These key components might include the following:

- A general description of the HAZCOM program
- A list of chemical components (MSD sheets) that may be present on-site
- General training requirements
- Frequency of required training
- General guidelines of personnel requirements of who will require training
- Other key components of HAZCOM

So the general safety program or safety manual will discuss the necessity and general applicability of HAZCOM to the company, and the site-specific safety plan will get into the details. These details might include the following:

- A specific list of chemicals (MSD sheets) that are expected to be present on-site when the work or task begins
- Specific training requirements detailing who will be trained, who will perform the training, what the training will consist of, and the length of time the training will take
- Frequency of site-specific training
- New product or newly introduced product training
- The specifics of where and how the training will be performed
- Other pertinent specifics regarding HAZCOM

Let's look at an example of when the safety program and site-specific safety plan might be the same document.

Example: An All-in-One Safety Plan

> If your company is small and never changing, the safety program (safety manual) and site-specific safety plan might be the same document. Let's go back to our hypothetical widget manufacturer again. This company has been making the same widget using the same machines and the same chemicals since the turn of the 20th century. All of their employees have more than 50 years of seniority, and all of their skills have been passed on, starting with their great-grandfathers, from generation to generation.
>
> The safety program (or safety manual) was originally written in 1920 and has never been revised except to change the names of different responsible persons. These revisions only needed to be performed every 20 years. The process never changes, the chemical used in the process never changes, the machinery never changes, and so on.
>
> If this example describes your company, then it is likely that the safety program and site-specific safety plan will be one and the same. It is also likely that your company experiences none, or at least very few, accidents. If your organization does not experience any accidents, it is unlikely that the reader would be interested in this material anyway; however, if your company does not fit this description, then you probably need both a safety program (general safety manual) and a site-specific safety plan as described earlier in this chapter.

Most organizations establish a goal of 100 percent field compliance. This 100 percent goal is for in-house and outside standards. I believe that companies have more success complying with their in-house standards rather than outside compliance issues. It is difficult to determine OSHA compliance because compliance officers may examine differently than in-house safety persons. Let's look at an example.

Example: When OSHA Comes Calling

> While I was working for a midsize metals manufacturer as their safety manager, OSHA showed up unannounced. The OSHA compliance officers informed management that they were on the

premises to investigate a complaint. The general consensus of the management team was that the company had an excellent in-house safety program that included an extensive inspection and compliance program. The accident incidence rate was low, and there was little doubt that we would pass this inspection with flying colors.

After examining the accident records and spending a couple of hours in the plant, the compliance officers announced that they were expanding the scope to a wall-to-wall inspection. Once the wall-to-wall inspection had been completed, OSHA issued citations. I was a bit disappointed with the issuance of the citations, but many members of the management team just could not believe it. It did not seem possible that we could pass our in-house inspections with flying colors but did not come away from the OSHA inspection without citations being issued. After all, we had a good safety program. We had numerous site-specific procedures. We followed the plan and the procedures, so one would think we would not receive any citations.

In retrospect, our in-house inspectors were nowhere near as thorough as the OSHA inspectors. These outside inspectors asked questions that the in-house inspectors would never have even thought of, and the outside inspectors saw things that the in-house inspectors saw every day and glanced right over. The point is that you can have a good safety program and a good site-specific safety plan and still end up with gaps in the field. *Note*: We will talk more extensively about attaining field compliance in the next chapter.

SUMMARY

Many companies have concerns with field safety compliance issues. These compliance concerns typically stem from outside sources, such as governmental agencies, or from sources within the company. No matter whether the compliance concerns are from internal or external sources, the concerns need to be addressed. To begin planning for field compliance, companies with a well-developed safety culture use two main types of documents. The general or "big picture" document is called a safety program (or safety manual). General safety issues are addressed in a general way in a safety program document.

To address compliance issues in a more "hands-on" manner, a site-specific safety plan should be used. This site-specific plan fills in the details of how safety compliance will be met, including the actual roles and responsibilities of those team members who are responsible for compliance issues. The site-specific safety plan document can and probably will contain other documents, such as JSAs. In an atmosphere where the safety culture is well developed, these JSAs will contain somewhat intricate detail regarding how team members will be protected against every hazard that will be encountered during work task activities.

Although various examples are offered, it is unlikely that any business can and should operate without both a safety program and a site-specific safety plan. The safety program and site-specific safety plan are two types of documents that are interrelated but serve different purposes and functions.

11
Safety Culture Barometers

All companies have a safety culture in place. This is the case even if the company in question has the worst safety record in existence. The current safety culture may be in need of improvement, but it does exist. If one plans to develop safety culture and increase the safety program's effectiveness, the current condition and state of the safety culture should be determined. Using the current state of the safety culture as a starting point, a route can be determined that can take the program from the current state and point it toward a zero-accidents culture. Unfortunately, it is sometimes difficult to determine an exact current location. When trying to determine this current location, the best that can sometimes be obtained is the general neighborhood. Some methods that can be used for this location determination, along with some of the obstacles that can be expected in making this determination, are discussed in this chapter.

A method that has been used with some measure of success to determine at least what neighborhood the state of a safety culture is in is accident trend analysis. If the causes of accident occurrences can be determined, and those factors that caused the occurrence are avoided in the future, then the likelihood that similar accidents will occur should be minimized. So, the strategic development of safety culture can and will lead to elimination of accident occurrences.

In principle, the trend analysis should be a relatively simple task. All that one needs to do is go back into the records for a predetermined amount of time—let's say three years—and look at the accident records for those three years. Three years is probably an average or at least a good point in time to consider. If there are lots of accident data to look through, the time frame could be cut down to one or two years. If the amount of accident data is scarce, the time frame can be increased to up to five years. If the operation has not changed much over the past seven or eight years and accident data are scarce (not untypical in a small company), the time frame studied could be increased to seven years or even to the last major change in operations.

Once past accident data are gathered, they need to be analyzed. Putting the accident data in a graphical format should help the reviewer determine key trends. These trends may give the reviewer some useful information. If the trend for accident occurrences is generally downward, then the reviewer may be able to determine what the company has done to promote this trend. Going a step further, if that downward trend in accidents is determined to be from (or thought to be from) a program or system that the company has instituted in a certain time frame, then maybe this program or system should be continued or enhanced to ensure that the trend of improvement continues in this direction (maybe even at a faster pace).

Accidents will not occur, or will at least be minimized, if our safety culture is strong. We need to determine what our weaknesses are; we can then formulate a plan to strengthen these weaknesses. In this chapter, we concentrate on determining what the weaknesses are. Formulating an objective is not necessary. Everyone already knows the objective, and everyone's objective should be the same: zero accidents, zero incidents, zero losses, and so on. Everyone should be striving for perfection. Accidents just *cannot* be accepted.

As mentioned earlier, determining an accident trend analysis should be relatively easy. In principle, at least, it is easy, or it should be; however, a variety of obstacles will probably be encountered when trying to determine the current status of safety culture. In the initial chapters, we discussed how an OSHA log can be a guide to potential problem areas. OSHA logs, accident reports, incident investigations, accident review boards, medical records, worker interviews, OSHA citations, and other information can provide a lot of information about safety culture development.

A problem that typically arises is that detailed records cannot be located for much of the information mentioned. Possibly the information needed to perform the assessment is locked away in someone's files. Possibly the information has been misfiled. Or even more likely, the information you are interested in never existed or was disposed of long ago.

OSHA LOGS AND OTHER ACCIDENT-RELATED INFORMATION

Various pitfalls can be encountered when attempting to use old OSHA logs in an effort to compare recordability. I am not saying that the OSHA logs should not be used. On the contrary, I believe that they should be used. I even equate the OSHA logs to a road map on several different occasions, but OSHA logs—just like old road maps—can be misleading. They may contain less than full disclosure or not include certain information that you as a reviewer might find pertinent. Or, as mentioned in earlier chapters, the OSHA log may contain overreported or improperly classified accidents.

Besides examining the OSHA 200 or 300 logs, the reviewer should consider obtaining the medical records that correspond to each accident. The diagnosis from the physician and/or caregiver should be used, along with the information from the accident report. Any accident investigation, pictures, interviews, and follow-up should also be compared to the accident report and medical treatment records. The reviewer may find that some information just does not add up. Because this subject has been discussed at various times in previous chapters of this book, I will not go into detail about it again in this chapter. Even though in principle the guidelines have been set up so that everyone reports injuries throughout the country in the same manner, you will probably still find considerable variation in how these rules are interpreted.

It is best to rely on accident records that are fresh and that can be personally verified. The older the accident occurrence is, the more likely it is that the recollection of anyone involved has been dulled. Also, it is less likely that all of the information is still intact or in the file. If other litigation surrounded an accident occurrence, do not be surprised that someone lifted out certain file information for some useful purpose but failed to bring it back. The older the case is, the less likely it is that available information will be complete and accurate.

Current trends are important but can be misleading. If you try to look at the present or very near past, you can become overconfident because of a lucky streak of no accidents or possibly late reporting. This overconfidence can results in statements like: "Gee, we haven't had an accident in over a month now. Our safety program must have really kicked in!"

This may be the case, and I hope the person making a statement like this turns out to be correct. Not having an accident in a month may be a great milestone for some companies. For other companies, one month without an accident might be considered failure. At minimum, I have found it best to look at accident trends over a six-month time frame. Trying to compare numbers in less than a six-month time frame can be a mistake. For many smaller companies, the accident data that a reviewer will gather over this time frame may be miniscule. Even with midsize companies, the six-month time frame might not provide enough data.

I have had numerous close associates who get overly excited about a milestone and begin to claim complete victory. These same people seem to get a bit too discouraged when multiple accident occurrences get reported over a short period. Hopefully, the multiple accident occurrences are more of a blip or anomaly or are more coincidental than due to a worsening culture.

TREND DETERMINATION

Once the challenge of locating accurate and complete records is met, the task of trend determination can begin. One of these key trends might be the accident incidence rate both for lost workday and recordable incidents. Another key trend that could probably be determined is the Experience Modification Rate (EMR). One should be able to obtain this number from the workers' compensation carrier. The workers' compensation carrier probably has already made a determination of which claims the most money has been expended on, which claims included the most days away from work, which parts of the body were affected most of the time, and so on. Getting information from the workers' compensation carrier can be another good source of information.

From this trend analysis, you might be able to pinpoint areas that seem to be begging for improvement. Different clients that I have been involved with had tremendously different accident trends. For instance, I worked with a chemical manufacturer at which 40 percent of the accidents — during a relatively short period time — were experienced by people with less than one year of seniority. This group consisted of about 10 percent of the workforce. These numbers seemed to indicate that an improvement was needed in the orientation, training, and mentoring of new hire workers.

Back injuries usually top the list for most costly body part injured, but I have had clients whose most costly injuries were from repetitive motion and not back injuries. These numbers seemed to indicate that an improvement was needed in ergonomic assessments. This finding eventually led to instituting extensive training, purchasing ergonomic equipment, making certain changes in operations, and more.

Vehicle accidents seem to continue to be a problem for most businesses whose team members spend time driving. I have had clients who experienced that accidents caused while a vehicle was backing up constitute more than 30 percent of total vehicular accidents. The point is that determining trends can give the safety person some ideas for improvement plans.

Looking at past accidents can give other insight. Find out if accident investigations have historically taken place after every accident occurred. If so, what was the quality of the accident investigations? Did the result of the investigations result in any recommendations? If so, were the recommendations followed? What group followed up to ensure that the recommendations were fully completed or implemented? If you are working with an underdeveloped safety culture, the answers to one or more of these questions will be no. I don't believe that this is uncommon. Many safety professionals have tried to analyze accident trends but postponed the task after finding that there is a lack of detailed information.

One of the companies that I worked with perceived that they had a problem with vehicular accidents. Various safety management personnel decided to profile a typical motor vehicle incident. They obtained vehicular accident files for a two-year period and examined the following parameters:

- Job title
- Age of driver
- Length of service with the company
- Object that had been collided with
- Vehicle direction
- Time of accident occurrence
- Location
- Type of road
- Type of vehicle
- Main cause of accident

Records for more than 30 vehicular accidents were included in this study. Initially, the study was to include 100 vehicles; however, after reviewing the reports, it was determined that many of the reports were filled out incompletely and did not contain adequate information to contribute to the study.

Detailed Records Must Be Kept

This point is very important. As I mentioned in Chapter 7, "Training," you can never have too much communication. Similarly, allow me to make another very important point: *You can never have too much documentation.* When reports are filed, the person accepting the reports must have an eye for detail. Incomplete reports cannot be accepted. The numbers that one uses to analyze trends are usually already so minimal that to not consider one case file because it is incomplete can significantly skew your results.

The time of day seemed to have some significance regarding vehicle accident occurrences. For this study it appeared that more than half of the accidents occurred between 0600 and 1200 hours. This time-of-occurrence statistic was more closely examined, and it was found that more than half of the morning accidents occurred between 0700 and 0930. Unfortunately, not enough information was available to determine why this would occur. Some people theorized that more driving is done during those hours than at any other time. But the truth be known, we just do not know. And probably we will never know. Here are some of the questions these issues raised:

- What exactly were those team members doing when they were involved in an accident?
- How many hours did the people work before being involved in an accident?
- What exactly were those team members assigned to do or supposed to be doing when they were involved in an accident?
- How many years of driving experience did the drivers have when they were involved in an accident?
- Had the drivers attended a safe driving course before their involvement in an accident?

The questions kept coming, but the answers for most of our questions were "We do not know." It is truly difficult to come up with the answer for why an accident occurred at a certain time when lots of information is not available to the person performing the analysis. Those persons performing analysis need to be careful with statistics. Sometimes analysts can give numbers a meaning and significance that is not warranted.

Example: Statistics Aren't Always What They Seem

As an example, I was a young driver when the National Safety Council seemed to blitz the media about safe driving. Much of the media blitz seemed to focus on near-peak holiday travel season or summer holiday and vacation season. The significant statistic that seemed to be a "motto" and that has been embedded in my memory for more than 30 years is that: "More than 80 percent of accidents occur within 50 miles of the driver's residence."

This statistical information was presented along with deaths on the highway during that same holiday period in the past year. Also, predictions of how many would die in the current year were mentioned. This information seemed to allude to how much more dangerous driving was done near your own home. One comedian used this as part of his routine. The punch line was: "I know how to avoid accidents—I'll move!"

The question that I have is: "What is the percentage of driving that one does within 50 miles of one's residence?" I don't know the answer, but for me and most of the people with whom I associate, I think the answer is probably near 80 percent. If this were the case, it would stand to reason that 80 percent of the accidents occurred within a 50-mile radius of someone's residence. This statistical information just does not seem to have the significance that the media blitz seemed to allude to.

I have enjoyed some definitive results in many of my attempts to perform accident trend analysis. I have also had many disappointments. In general, my disappointment arises when I discover that my attempt to analyze trends is overshadowed by a lack of data. I have found that in many instances there typically is not enough information to form a clear conclusion. The sample base and database are just not large enough.

Example: Back Injuries

In my earlier days as a safety professional at a manufacturing plant, the biggest concern to management was back injuries. This company made a valiant effort to eliminate any heavy lifting during any assigned work tasks. One back injury was enough to send

(continued)

workers' compensation premiums through the roof, and management wanted a thorough analysis of what was causing these back injuries and what could be done to prevent them. Extensive training programs for proper lifting were repeated several times per year, and the topic was constantly the subject of the daily safety meeting.

Looking over the back injuries that had occurred over the past 10 years at this company elicited some interesting results. One of the results was that some of the most costly back injury claims stemmed from arthritic backs and degenerative disk disease. Records indicated that these workers had little to no heavy lifting during their working careers. Their medical files indicated that their disorders were most likely caused by aging or hereditary reasons. My analysis of this situation was that although the company looked at these back injuries as a loss and huge expense, the losses they were incurring were due more from outside influences than work activity; however, I believed at the time (and continue to believe) that the investment this company made in safe lifting techniques, safety talks, and other training was paying off and would continue to pay off in the long run.

Accurate measurements must be taken and recorded. As mentioned with detailed information and incomplete reporting, the accuracy of any measurements is important. Besides taking physical measurements, I recommend taking pictures. I think it is a good idea for all supervisors to have a camera immediately available. A digital camera would be excellent, but a disposable camera will usually suffice.

Not that anyone would skew results on purpose, but I have reviewed accident reports and followed up with victim interviews that did not appear to match. This type of situation is not a one-time occurrence. It is sometimes difficult to discern if the people I interviewed had lapses or if the written report was inaccurate. If pictures are available, I have found them to be helpful when accident occurrences need to be re-created.

SAFETY AND HEALTH PROGRAM ANALYSIS METHODS

Clearly, determining the current state of the safety culture is not necessarily an easy job. In fact, determining the state of the safety culture can be a difficult job. I believe the job of culture assessment gets harder to do as the number of accident cases that you can examine decreases. So, if you have a very small number of accidents in your current culture, perhaps a hearty congratulations is in order. Or perhaps the management realizes that even though accident data are scarce, an improvement and development of culture is necessary. If the accident data are not sufficient to point out deficiencies, alternate methods need to be utilized. One method that comes to mind has been developed by OSHA.

As an alternative to studying accident cases, OSHA has come up with a method to analyze weakness in what they call the Safety and Health Program. I believe that what OSHA refers to as a Safety and Health Program Assessment could also be considered a safety culture assessment. I have found much of this information to be useful, but I must warn readers in advance that OSHA has set the bar for excellence quite high. Be advised that in some cases reaching the top level may be nearly impossible. So, do not be too disappointed if you find a variety of sections where the survey indicates a need for improvement.

Only highlights from this assessment have been included. These highlights show only the answers determined by OSHA to be the best or optimum answer. OSHA has included a worksheet with a detailed grading system so that interested parties can assess their safety and health program. OSHA groups keys to success in various categories. For instance, "Management Leadership and Employee Participation" is the first category that I list from the Safety and Health Program Assessment Worksheet. The first section of this category discusses safety and health policy. OSHA's category is entitled "Clear worksite safety and health policy." Readers then get a chance to rate their own programs.

If the program being rated answers each category with the top answer, this would indicate a well-developed safety culture. If the answers given are other than the top answer, this would indicate that there is room for improvement. For the worksite safety and health policy category, the top answer is: "There is a clear worksite safety and health policy that employees can accept, can explain, and fully understand." If the reader answers in this way, it would indicate a well-developed safety culture for that category. I have included all of the categories offered by OSHA along with all of the top answers.

Management Leadership and Employee Participation

- *Clear worksite safety and health policy*: There is a clear worksite safety and health policy that employees can accept, can explain, and fully understand.

- *Clear goals and objectives are set and communicated*: All employees are involved in developing goals and can explain desired results and how results are measured.
- *Management leadership*: All employees can gives examples of management's commitment to safety and health.
- *Management example*: All employees recognize that management always follows the rules and addresses the safety behavior of others.
- *Employee involvement*: All employees have ownership of safety and health and can explain their roles.
- *Assigned safety and health responsibilities*: All employees can explain what performance is expected of them.
- *Authority and resources for safety and health*: All employees believe they have the necessary authority and resources to meet their responsibilities.
- *Accountability*: Personnel are held accountable and all performance is addressed with appropriate consequences.
- *Program review (quality assurance)*: In addition to a comprehensive review, a process is used that drives continuous correction.

Worksite Analysis

- *Hazard identification (expert survey)*: Comprehensive expert surveys are conducted regularly and result in corrective action and updated hazard inventories.
- *Hazard identification (change analysis)*: Every planned or new facility, process, material, or equipment is fully reviewed by a competent team, along with affected workers.
- *Hazard identification (job and process analysis)*: A current hazard analysis exists for all jobs, processes, and material; it is understood by all employees; and employees have had input into the analysis for their jobs.
- *Hazard identification (inspection)*: Employees and supervisors are trained, conduct routine joint inspections, and all items are corrected.
- *Hazard reporting system*: A system exists for hazard reporting, employees feel comfortable using it, and employees feel comfortable correcting hazards on their own initiative.
- *Accident/incident investigation*: All loss-producing incidents and near-misses are investigated for root cause with effective prevention.
- *Injury/illnesses analysis*: Data trends are fully analyzed and displayed, common causes are communicated, management ensures prevention, and employees are fully aware of trends, causes, and means of prevention.

Hazard Prevention and Control

- *Timely and effective hazard control*: Hazard controls are fully in place, known, and supported by the workforce, with concentration on engineering controls and safe work procedures.
- *Facility and equipment maintenance*: Operators are trained to recognize maintenance needs and perform and order maintenance on schedule.
- *Emergency planning and preparation*: There is an effective emergency response plan and employees know immediately how to respond as a result of effective planning, training, and drills.
- *Emergency equipment*: Facility is fully equipped for emergencies, all systems and equipment are in place and regularly tested, and all personnel know how to use equipment and communicate during emergencies.
- *Medical program (health providers)*: Occupational health providers are regularly on-site and fully involved.
- *Medical program (emergency care)*: Personnel who are fully trained in emergency medicine are always available on-site.

Safety and Health Training

- *Employees learn hazards (how to protect themselves and others)*: Facility is committed to high-quality employee hazard training, ensures that all participate, and provides regular updates; in addition, employees can demonstrate proficiency in, and support of, all areas covered by training.
- *Supervisors learn responsibilities and underlying reasons*: All supervisors assist in worksite hazard analysis, ensure physical protections, reinforce training, enforce discipline, and can explain work procedures based on the training provided to them.
- *Managers learn safety and health program management*: All managers have received formal training in safety and health management responsibilities.

The complete listing of this self-assessment can be found on the Internet at www.osha-slc.gov/SLTC/safetyhealth_ecat/asmnt_worksheet.htm. If the reader has not answered in the indicated manner, there is at least some room for improvement in the development of safety culture. Looking at the recommended answers, it appears that OSHA has set the bar high for safety. To get a better feel for how OSHA would rate a safety program, all you need to do is complete the whole survey and rate yourself.

PERCEIVED OBSTACLES TO SAFETY

Besides the self-assessment mentioned, OSHA has published a list of obstacles to a successful Safety and Health Program on the same link as mentioned previously. In this section, they list the supervisor and employee viewpoints of what the obstacles are in preventing progress in the development of an improved safety culture. Although the list segregates the views of the supervisors and employees, I noticed that many of the points are the same or are at least very similar. The obstacles that OSHA has listed are extensive. I have taken OSHA's list and condensed it to what I believe are the key obstacles for both the workers and supervisors.

Take the time to completely look over this list. If you are under the impression that supervisory staff and hourly staff have viewpoints that differ substantially from one another, I think the findings will be surprising. From what OSHA has published, it appears that in a lot of respects the concerns about obstacles to an effective safety program from both the workers' and supervisors' perspectives are, for the most part, shared concerns. A partial list of these shared concerns is as follows:

Supervisor- and Employee-Identified Obstacles

- Fear of losing my job
- No money for needed changes
- No support from upper management
- Lack of trust: poor ethics within organization
- Lack of open communication and listening
- Competing priorities: production is number one

In the following sections, I discuss each of these obstacles and describe how they can affect the development of safety culture.

SUPERVISOR- AND EMPLOYEE-IDENTIFIED OBSTACLES

Fear of Losing My Job

Even if you have a great love for your current position and are working for a stable company in a stable career path, almost everyone has had thoughts from time to time about losing his or her job. A certain amount of fear regarding the potential for a job loss is probably considered normal. After all, for most people their jobs are how they support their lifestyles and pay their bills. Having concern over job security is not an unreasonable concern. Having a concern about being injured on the job is also not unreasonable; however, if team members believe that the current culture they are working in uses fear and intimidation as a means to motivate workers, it will probably take a considerable amount of time to change those team members' points of view.

If team members believe that they could not voice their legitimate concerns about safety problems that they encounter on the job without fear of reprisal, then that belief needs to be changed immediately. An attitude or perception that complaints from team members regarding safe and healthful working conditions cause negative effects toward long-term successful employment will impede safety culture development.

Management must strive to reassure people by both word and action that any team member's legitimate concern regarding safety is respected by management and that those team members who bring up legitimate safety concerns will not be punished or lose their jobs.

No Money for Needed Changes

No money for needed changes is a typical battle cry. Even in our personal households, there is typically not enough money available to make the purchases we would like to make. Personal households have a budget similar to businesses having a budget. Sometimes there is just no money around to do the things we would like to do. Consider the analogy of buying a new car. If we are currently driving a car, we probably have a plan as far as how many years and how many miles we would like to keep this car.

In my personal household, I have been trying to replace my personal vehicles or at least seriously reassess them when they reach or are approaching 100,000 miles. When family members ask me when we can get a new car, I advise them that we are not yet in the market for a new car and give them some parameters of when we might be willing to discuss the purchase; however, sometimes repairs to a vehicle can be very costly, but the vehicle might not be worth the cost of repairs. So, I revisited my budget and made some adjustments. I decided that instead of spending a lot of money on repairing the old vehicle, I put the repair money toward the purchase of a new vehicle.

Most companies operate in a similar fashion. Product lines are expected to last a certain amount of time. Depreciation becomes a factor, and the costs of repairs, maintenance, and downtime also become factors. Team members, supervisors, and management may all agree that a certain product line presents hazards and should be replaced. A disagreement may surface when the accountants of the world advise that there is just no money available for the replacement. Sometimes no matter what kind of argument is presented, the answer is just NO!

Example: The Story of Chitty-Chitty Bang-Bang

> A chemical distributor purchased its fleet of heavy-duty, diesel-powered tractors from General Motors. At one branch location, this company had a fleet of six tractors. This company had an informal plan of replacing these tractors at the rate of approximately one per year. When I first was hired, I noticed that we were renting one tractor rather than using one from our own fleet. I questioned why we would rent a tractor rather than lease or own. I was informed that we did own another tractor, but that it was in for major repair. After further discussion, I found out that this tractor *always* seemed to be in for repair. It was in for repair so much that it had gained the nickname of Chitty-Chitty Bang-Bang, which was the noise that it usually emitted before it was taken in for service.
>
> I obtained a variety of background information regarding this "lemon" of a tractor from a variety of sources. I talked to all of the drivers who worked at the branch. I talked to the warehouse person regarding this tractor. I talked to the service technicians and the service manager at the GM dealership that provided service on this and all of our tractors. I also talked to operations management, purchasing, and corporate management. It appears that the purchase of Chitty-Chitty Bang-Bang was an experiment. Our purchasing department was interested in buying new equipment that was fuel efficient and could be easily maintained. Chitty-Chitty Bang-Bang was supposed to fit that bill. It was a gasoline-powered unit that was geared with a 15-speed transmission (5 by 3). The engine was supposed to burn clean, last a long time, and be easily replaced when—if and when—it finally wore out.
>
> The experience with this tractor was not good. The engine was just not powerful enough to pull the loads that we demanded, especially over the long haul. No driver wanted to drive this tractor. The repair people were reluctant to repair it, and breakdown after breakdown continued. If ever there was any such piece of equipment that could be called a lemon, this was it. Drivers complained that it was

a safety hazard because it would break down at awkward locations and times, putting the driver's safety at risk. The manufacturer had cut out producing the unit due to the tremendous complaints and service problems. As this piece of equipment got older, parts and repairs became more difficult to obtain. And the best alternative became the rental tractor.

You might ask, "Why not just buy a new one?" The answer was easy. The accountants just said NO! No explanation would change their minds. We were stuck with Chitty-Chitty Bang-Bang until further notice. Business took a downturn, and even though the tractor was completely depreciated, we would not be getting any new replacements for at least another year.

This problem just did not go away. The operations people did not want to pay for a rental tractor and continue to repair Chitty-Chitty Bang-Bang. Even though Chitty-Chitty Bang-Bang was not being operated, there were still expenses associated with it. These expenses included maintenance, license plates, stickers, insurance, and storage fees. We were stuck. We could not sell it or give it away, yet we had to keep maintaining it and still paying for a rental.

Every month the subject of Chitty-Chitty Bang-Bang came up and was reviewed. No one appeared to be satisfied with a year's wait. After about six months, our corporate people informed the branch management that our request for a new tractor would be granted and that Chitty-Chitty Bang-Bang was going to be taken off the books. When this new tractor arrived, it was a welcome sight. Seeing Chitty-Chitty Bang-Bang drive off the property for the last time was an even more welcome sight. The successful end to this problem added to the team members' faith in management and to safety culture development in general.

No Support from Upper Management

No matter how supportive upper management tries to become, there is always room to be more supportive. This may include increased support for the safety program with money, time, effort, and/or attention. Upper management is typically pulled in many different directions at once. Management tries to coordinate efforts to ensure that the end product of the business is better, safer, cheaper, and faster than the competition. With this competitive marketplace, there is only so much support to go around. Every department, including the safety department, is always looking for more support from upper management. These other departments looking for support from upper management may include the following:

- Operations
- Maintenance
- Engineering
- Human resources
- Quality assurance/quality control
- Security
- And literally every department within the company

With larger companies, getting support from upper management appears to resemble lobbying efforts. Sometimes the squeaky wheel or the well-connected seem to get the grease. These are life's realities, and they need to be taken into consideration. As mentioned with the example of Chitty-Chitty Bang-Bang, it took a lot of effort over a long time to address a problem that was relatively simple to understand and easy to fix. Strength of character and persistence are needed to resolve some seemingly simple problems. If you stay positive and stick with it, your chances for success increase.

Lack of Trust: Poor Ethics within Organization

Certain people will never completely trust their co-workers, management, or both. Sometimes this lack of trust is warranted, but sometimes it is overblown. I have a good friend who tells me quite often that "I trust my fellow card players are honest, but I still want the opportunity to cut the cards."

It appears that the management of most companies makes an honest effort to hire people that they believe are and will continue to be "good corporate citizens." The attempts to hire the best people possible for the available positions sometimes force the decision makers to compromise. Ethics within the organization can be affected from top to bottom. I believe that when the top management in the organization is making the effort to run the business in an ethical manner, this becomes obvious to everyone at the firm. When upper management is running the business in a manner that is considered "shady," this also becomes obvious and can have a negative effect.

Lack of Open Communication and Listening

Failure to adequately communicate is a universal problem. As mentioned before, you can never have too much communication. As an educator, I learned that I was lucky to get my point across to 80 percent of my students. The material that I would use was meant to reach approximately 60 percent of my class. Typically, 20 percent of the students were disinterested or below the level I was teaching at, and 20 percent already had an in-depth knowledge of the subject and material.

I believe that communication is always a two-way street. When communication becomes a one-way street, it should no longer be termed communication. If communication is only one way, it would be better termed an edict, order, or directive, but not communication. Effective communicators are typically not only good talkers but also good listeners.

Did you notice the word "open" in front of communication? I get the impression that the people who came up with this obstacle felt that they could not mention any safety deficiencies at an "open" meeting. I have met many people who would never say anything in the open meeting format. Some team members do not wish to speak in front of any type of group. Some people just don't think it is their "place" to speak in front of the group. These same people would make a point to discuss things with you on a one-on-one basis.

Management should attempt to allow for reasonable input from team members at group meetings. I think it is helpful when management says at a group meeting that they have been advised of a certain subject in one-on-one meetings and shares the situation at a group meeting. Although I have been at meetings where communication appears to be one way, at least management gave everyone the opportunity to speak, even though no one exercised their right to do so. What I have found offensive is when management never holds group meetings and thereby no one has the opportunity to speak or communicate.

Competing Priorities: Production Is Number One

This "production is number one" obstacle is the oldest obstacle I know of. I have heard it ever since I started to work. The product that any firm produces is its life blood. Without the production of the end product, any business will soon cease to exist. Is production number one? My answer is a simple YES. Production *is* number one. But when we get past the simple YES, we see that there are a whole lot of number ones. These other number ones include the following:

- Safety
- Quality
- Engineering
- Maintenance
- Research
- And the list goes on and on.

Every one of the departments and all of the people within the departments have an important purpose. If we remember the Hawthorne Effect, we should be treating all of our team members as if they are important cogs on the wheel—because they are.

Example: The Friendly Skies

> After September 11, 2001, I was riding on a Southwest Airlines flight as a standby passenger. After waiting in long security lines and several flight changes, I was finally allowed to board my flight. For those of you who have never flown Southwest, they do not have a first-class cabin like many other airlines. They board all customers on a first-come, first-served basis in groups of 30.
>
> As it turns out, I was in the third group of 30 passengers to be called for boarding. As you can imagine, the plane was quite full when I stepped aboard. As I got onto the plane, the flight attendant smiled and asked how I was doing that evening. I jokingly responded that I was doing fine and that I held a first-class ticket and was wondering where the first-class cabin was. A big smile came to her face as she told me just to sit anywhere that I wanted, and that all of the seats on the airplane were first class!
>
> Many companies pride themselves on making their places of employment the best places to work as rated by their own employees. I believe that when a company makes the commitment to the team members, the team members will make a commitment to the company.

OSHA typically focuses on safety culture from the workers' viewpoint rather than from the management's viewpoint. There is a good reason for this approach. After all, the CEOs of the world are rarely found contributing to the company's accident record. As far as I know, OSHA will probably not be interviewing the top management to determine what they believe the obstacles are in development of safety culture.

POTENTIAL CEO AND TOP MANAGEMENT-IDENTIFIED OBSTACLES

I have put myself in the place of the CEO and have surmised what the CEO likely believes are the obstacles:

- Fear of losing my job
- No money for needed changes
- No support from the board of directors
- Lack of trust: poor ethics within organization
- Lack of open communication and listening
- Competing priorities: production is number one

It is amazing that my surmised list looks so much like the list that the team members and supervisors formed. CEOs always seem to be losing their jobs. The board members on the board of directors are typically conservative by nature. A CEO who thinks *too far* out of the box will probably be on a slippery surface. Fear of losing that top job is certainly a real fear for the CEO.

Money is always an obstacle. No matter how much money you make and spend, there never seems to be enough of it. Not having the money for needed changes is probably a real obstacle for the CEO. No support from the board of directors is probably something that a CEO needs to be concerned about. Watchful eyes of board members are usually on budgets and purse strings.

A lack of trust among the CEO, board members, independent auditors, shareholder concerns, and employee concerns can affect the CEO. The lack of communication and listening become apparent especially when an organization is struggling financially. There is a lack of patience and general mistrust that top management must concern themselves with.

Production is number one; however, I do not view this as a conflicting priority. We all realize that if the business ceases to make the product or service that it sells, it will no longer exist. All of our company core values are number one. This includes safety, quality, engineering, maintenance, and so on.

As you read the potential obstacles facing the top management in regard to safety culture, perhaps you realized as I did that safety is a shared goal. In simplistic terms, accidents are just a "bad" thing. They are bad for the team members because they suffer physical pain and loss of functioning in a normal manner; they are bad for supervisors, if for no other reason than because life becomes more challenging when a valued team member must be replaced; and they are bad for the top management people who hopefully realize that accidents, besides causing needless pain, suffering losses in production, and so on, cause the costs of producing their product to increase. This increase in costs reduces profits and decreases the value of the company. Within the different entities of a company—whether these entities are hourly production workers or members of the board of directors—no one within any organization wants an underdeveloped safety culture. No one wants an accident. Besides being painful, accidents are also bad business.

SUMMARY

This chapter discussed how to determine the status of safety culture. One of the methods discussed was accident trend analysis. Accident trends can help determine areas within an existing culture that are in need of bolstering. Although government regulations are in place to ensure that everyone reports and records injuries in the same manner, life's experiences have determined that variations exist in reporting that span from both extremes. These extremes include both overreporting and underreporting of injuries and illnesses. Recent accident data that are complete and verifiable should be used whenever possible. Older accident reports that are incomplete and that contain references to team members who are no longer with the company or that are difficult to find and verify should be used for trend analysis only with extreme caution.

Besides the use of accident trend analysis, OSHA has introduced a self-check assessment that can help determine areas of weakness within a safety culture. This assessment goes through what OSHA believes are the key components of a Safety Program. I believe that it would be difficult for a company to complete the self-assessment and find that it did not show the need for substantial improvement.

At the end of the chapter, a discussion on perceived obstacles was included. OSHA published a study based on interview results taken from supervisors and hourly workers, who were asked to list what they, as individuals, believed were the biggest obstacles to developing an effective Safety Program. Highlights from this OSHA study are cited, and a correlation is made regarding how many of the obstacles pointed out by the hourly workers were actually shared by the supervisors. Potential obstacles as viewed from top management or the CEO of a typical organization are also discussed. It appears that safety truly is a shared value. Everybody at all levels should be concerned about safety.

12
Measuring Field Compliance

There are two commonly used and effective ways to measure field compliance: field safety inspections and hands-on field audits. Notice that the terms "field" and "hands-on" are used in the previous sentence. These are key words because trying to perform inspections and audits from the office or the "ivory tower" is just not the same as being there.

Even in the inspection and audit process, enhancing the Hawthorne Effect must be a priority. If management fails to spend adequate time in the field or on the shop floor, mixed signals are being sent to the workforce. In order to show people that they are important and that management cares about them, top executives should spend a considerable amount of time in the area where work is taking place. Management should be extremely observant of this ongoing work and in constant communication with some of the people who perform the work. This involvement of top management in safety issues is another key to success.

The president, CEO, or other top company official showing up and participating in a safety meeting shows that top management cares about safety. This message will enhance the Hawthorne Effect. If top management shows this same attention regarding the results of a safety inspection or safety audit, this message will also reinforce that the company believes that things such as worker safety and health are important. Management has a significant role in supporting safety culture. Management's support cannot end at the kickoff meeting but must be shown and reinforced every step of the way. This support must be shown when inspections and audits are taking place.

The inspection and audit process should be a positive rather than a negative experience. If your safety culture is developing, your inspections should show that shortcomings that were pointed out during previous inspections have been corrected and that the work process, along with the safety process, is continually improving.

Described as follows are two tools that are used extensively in measuring field compliance: the audit and the inspection. I define these two tools as related but different. The main differences between them are in their scope. Many health and safety professionals do not realize or believe that there is a difference between the safety program and the site-specific safety plan. Similarly, many health and safety professionals will not admit to a difference between the safety inspection and safety audit, but there is a difference—and the difference is substantial. Safety inspections, and why these safety inspections are important, is discussed in the following section. After completing the discussion on safety inspections, safety audits are discussed.

MEASURING FIELD COMPLIANCE USING SAFETY INSPECTIONS

Safety inspections are a valuable tool to aid in determining field compliance. Inspections can be used to reveal gaps between field activities and safety programs and plans. Safety inspections take place at various levels, including the following:

- Worker level
- Field supervisor level
- General foreperson level
- Safety representative level
- Other level

WORKER-LEVEL INSPECTIONS

Workers should be inspecting their areas regularly and making improvements as necessary or reporting deficiencies to their supervisors so that unsafe conditions are addressed. Having everyone involved with the safety program, including workers, is of utmost importance. Workers must realize that inspecting their work area for safety concerns, and correcting those concerns, is an important shared value for workers and management. Workers should be inspecting their work area for safety hazards numerous times during their shift.

The most important inspection would likely occur at the beginning of the shift before work inception. This inspection, whether formal and documented or informal and undocumented is very important. (*Note*: It is recommended that a formal, documented inspection take place at a minimum at the beginning of the shift.) Making sure that the work area is safe before the inception of work is a basic foundation for safe work throughout the day. Finding safety hazards and correcting them before work begins is a positive step for the company's safety culture and culture in general.

Housekeeping

Housekeeping is an important part of safety. Worker-level inspections should stress good housekeeping as a basic requirement. Good housekeeping can mean a lot of different issues to a lot of different people. It can mean keeping the shop floor free from debris and scrap. Maybe the floor needs to be cleaned, mopped, or swept once per shift. Maybe it means making your work area free from slip, trip, and fall hazards. Maybe it means making sure that used gloves and work items are promptly disposed of in the proper manner. Maybe it means the oil cans and excess grease have been put back into proper storage.

Pretrip Inspections

For workers who are tasked to operate equipment, whether that equipment is heavy equipment on a construction site or presses or equipment used in a manufacturing process, the worker needs to perform a prework inspection of this equipment. For workers who are asked to operate mobile equipment such as trucks, forklifts, Bobcats, skidloaders, and the like, a written pretrip inspection must take place. Although there are a variety of ways to document this pretrip inspection, I recommend that the inspection be recorded on a prenumbered pad containing pressure-sensitive carbons with at least the original and one copy.

Now that we are in the computer age, people have found ways to document a pretrip inspection on their computers when log-in is performed at the beginning of the shift. Customized questions pop up that the operator must answer before being allowed to continue using the computer. Computers are a great way to store data and keep records;

however, many businesses are not quite ready for computerized pretrip inspections.

An alternative to the high-tech documentation is the old-fashioned, low-tech documentation. This low-tech way of doing things might be a place in the garage area or office with clipboards corresponding to various pieces of equipment. On these clipboards are copies of pretrip inspection forms. Each day before the beginning of work, each operator finds the clipboard corresponding to the piece of equipment he or she will be operating and fills out the form. The old forms (from the previous day's activities) are placed at the bottom of the pile. Any deficiencies that are noted are discussed with the front-line supervisor. If there is any question regarding safe operating condition of any equipment, that piece of equipment is not operated until a group decision is made among the team member, supervisor, and others (possibly the maintenance and safety person).

At the end of the month (or other appropriate time frame), the supervisor gathers the forms and files them appropriately. Possibly the maintenance department keeps these records on file, or they could be kept anywhere that is appropriate. Exactly how the records are kept is certainly important, but what is even more important is how the pretrip inspection is performed. One can easily "fudge" an inspection form. Team members can put an "X" in any box indicating everything is fine without having to even look at the equipment. So whether you have a computerized system or the old-fashioned clipboard method, it is not the documentation of the inspection but the quality of the inspection itself that counts. An example of what this pretrip inspection form might look like is included in Figure 12–1. (*Note*: Figure 12–1 is not intended to be a working example of this form, but rather a template to give the form preparer ideas regarding the type of information that belongs on this form to fit a particular piece of equipment or operation.)

This form should be modified and customized for the equipment being used and the location where the equipment is being used. Some information is universal, and the principle behind utilization of this form is easy to understand. The operator is responsible for checking this piece of equipment for safe operation before using equipment. It is called a pretrip form because it needs to be filled out before taking this piece of equipment on a "trip." The universal information on this form includes the following:

- Unique equipment identifier
- Operator's name
- Date and time of the pretrip inspection

Pre-/Posttrip Heavy Equipment Check

Operator's Name: __ _____ _____
                        (Printed)   (Signed)

Date: _____ Time: ___ Equipment ID: _____

Check "Yes" if equipment "OK". Check "No" if equipment needs repair/adjustment. Any comments or description can be included in the space indicated below.

		YES	NO
1	Headlights		
2	Taillights		
3	Back-up alarm		
4	Seat/seatbelt		
5	Oil level/condition		
6	Coolant level		
7	Belts/hoses		
8	Transmission fluid level/condition		
9	Battery(ies) securement/fluid level		
10	Glass		
11	Guards		
12	Foot pedals		
13	Arm levers		
14	Windshield wipers		
15	Hydraulic fluid level/condition		
16	Brakes		
17	Fire extinguisher(s)		
18	Fuel system		
19	Exhaust system		
20	Gauges		
21	Keys/locks		
22	General running condition		
23	Tires/tracks		
24	Bucket		
25	Outriggers		
26	Steering/guidance system		
27	Charging system		
28	Leaks		

_____
_____
_____
_____
_____
_____
_____

Figure 12–1 Pretrip Inspection Form.

- General running condition
- Guards in place
- Observable leaks

Information that is more specific to certain pieces of equipment might include the condition of the following:

- Lighting/headlights/taillights
- Fire suppression system/fire extinguishers
- Seat/seatbelt
- Gauges
- And many more

Let's take a closer look at the universal information:

- *Unique equipment identifier.* The unique equipment identifier might be the name of the equipment or inventory number. An example of this identifier might be "Large CAT excavator #62654." Or "American Crane #53423." In both cases the supervisor knows the general type of equipment that is being referred to and which specific piece of equipment. Large CAT might just not do, and #62654 does not provide a lot of description, especially for team members who may not be as familiar with the equipment as the operators.
- *Operator's name.* The name of the operator and the operator's employee number or clock number is typically included on the pretrip form. The name should be clearly printed and the signature should be included. Anytime someone is asked to sign his or her name seems to make the document appear more official. This can be helpful for those who need to be reminded of how serious a pretrip inspection form really is.
- *Date and time of the pretrip inspection.* Documenting the date and time that a deficiency was noted can be helpful. If a later determination is made that reveals the problem persisted without being reported, this may indicate a need for further operator training, a lackadaisical attitude about the pretrip inspections, or other problems. If no action is taken after a deficiency is noted on the pretrip form, this may indicate poor supervision or lack of follow-up by maintenance or other problems.
- *General running condition.* General running condition being checked off as a concern indicates that the machine may just not be running properly but that a specific concern cannot be pinpointed. Some symptoms of what the machine does should be noted at the bottom of the form.

- *Guards in place.* All guards and safety equipment should be checked and in place before operating any machinery.
- *Observable leaks.* Any leaks observed need immediate attention. Major or minor leaks might be cause for immediate shutdown. Minor leaks that do not cause immediate shutdown should be contained with a drip pan or "diaper." (Actually, disposable diapers can be used quite often to contain small drips on industrial machinery that is experiencing oil leaks.) Besides the leaks potentially causing slip, trip, or fall hazards, the leaks might cause an environmental hazard or constitute a release.

The pretrip inspection form shown in Figure 12–1 is conceptual in nature and should not be used as a universal pretrip form. Instead, an in-house safety professional can use the concepts in Figure 12–1 to develop a customized pretrip form for each type of machine or operation being used. Also, besides the pretrip signoff, a signoff and formal inspection at the end of the shift should be considered. These pretrip and end-of-shift inspections can be critical when more than one person is assigned to any machine. There may be rare occasions when damage occurs to a machine or a facility that can be traced to an operator through pretrip inspection forms. Once is too often to discover damage obviously caused by equipment when no equipment operator admits to having caused the damage. This situation will typically cast aspersions over all team members when the discovered damage appears to have been caused by no one. (*Note*: We talk about integrity and safety culture more extensively in later chapters.)

Remember that some of the equipment operators may have limited computer literacy or written communications skills, so the written report and computer reports may be terse or somewhat primitive. The descriptions of problems may appear nebulous. Even though these team members may not excel at written communications, they typically excel at understanding how to operate equipment. And they typically can tell when equipment is not operating properly or is in need of maintenance. But let's get back to a typical pretrip inspection.

I prefer to have pretrip inspection documentation come in bound books or pads. These pads consist of pressure-sensitive paper of three different colors. The original should be white, and the copies should be yellow and pink. The white copy gets filled out daily by the operator and turned in at the beginning of each shift. Having the pretrip inspections in a bound book rather than a three-ring binder or making photocopies is important. The three-ring binder and the copies have a tendency to be misplaced or disappear. The bound books seem to have much less of an affinity to disappear or be misplaced. The department supervisor looks over the white copy to help determine the equipment's potential worthiness for work.

If you use the three-colored pad, the white copy could be for the department records, the yellow copy could be for maintenance, and the pink copy might stay bound in the book and remain with the equipment. Many pretrip inspection systems work equally well in different situations. Most, if not all, companies require that a pretrip inspection be documented on a daily basis. If your current pretrip inspection is effective, I am not recommending a change; however, if your safety culture is ripe for a change and your current pretrip inspection process needs an upgrade, consider using a system similar to the one discussed here.

In the three-color system, if there are no repairs or adjustments needed, the white copy is filed within the equipment's home department. The yellow copy is filed with maintenance and possibly collected during regular preventive maintenance checks. The pink copy stays in the book.

If the operator notes the need for adjustment or repair, the supervisor uses the pretrip inspection form to generate a work order for maintenance to initiate repairs. If the repair looks serious enough, the supervisor and operator will jointly make a decision about whether the equipment should be sidelined or if the equipment can safely operate in its current state. The judgment about whether equipment should continue to operate or be sidelined is too often a matter for debate.

A Closer Look at Professional Judgment

Let's look at a situation where an operator reports that the forklift or piece of mobile equipment is using oil. The first step in this scenario is that the team member performs a pretrip inspection and notices that the oil dipstick has indicated a shortage of oil. Consequently, the team member informs the supervisor that the equipment is using oil. At this point, a couple of questions arise: (1) How much oil is the equipment using? and (2) Where is the oil going? Hopefully, the operator is adding oil as necessary and required in accordance with the manufacturer's recommendations and established procedure.

When considering potential answers for these two questions, these answers can lead in a variety of directions. Consider the answer for the second question first. The oil can be going to one or more places. The oil might be getting used in the combustion process and burned. It might be leaking through a gasket or seal onto the exhaust manifold and/or hot surface and burned. The oil might be leaking from a sending unit, gasket, seal, hose coupling, or other place and be slowly dripping to the ground. Or the oil might be getting slowly spread throughout the warehouse at an unobservable rate. Or is the oil leaking profusely? Is the rate of the leak a quart every day, every hour, every minute, every month, or does it vary? Or maybe maintenance performed a preventive maintenance service on this equipment yesterday and failed to put in the required amount of oil when the technician refilled the crankcase.

Notice that in trying to answer the second question, the first question also needed to be answered. But what happens if the team member cannot answer the second question? Let's say that the team member cannot determine where the oil is going, but knows that the dipstick shows a shortage of oil on certain occasions. If the oil is missing but it does not seem to be leaking anywhere, this might mean that it is being burned as mentioned in the previous scenarios.

So then we need to revisit the first question again. We need to know how much oil is being used so that we can determine whether this equipment should be sidelined immediately or if it is safe to operate. If the consumption is slow and steady, there may be no immediate need to sideline this equipment. Maybe this piece of equipment is going to be sidelined for routine maintenance in the near future anyway. This problem can be noted and looked at during the routine maintenance check. Or maybe this equipment is dripping oil at a pretty good rate, leaving some slick spots throughout the warehouse and loading dock and in the truck beds of customers, causing potential slip, trip, and fall hazards behind wherever this piece of equipment has traveled.

The use of oil does not necessarily mean an immediate shutdown of work and sidelining of equipment. The whole picture needs to be examined, and a value judgment needs to be made. The question of urgency also needs to be considered. Is this truly an urgent problem? If after analysis of the problem the investigation indicates that the problem needs an immediate solution because of the hazard that continuing operation of the equipment will create, then the equipment must be sidelined; however, if the problem appears to be minor and not urgent and no imminent danger or serious hazard is involved, then a decision regarding an acceptable timetable for a fix needs to be established and agreed on.

A decision such as the one described for this situation will involve individual professional judgment and integrity. This can be related to everyday real life by examining how individuals treat their personal automobiles or vehicles. Some vehicle owners are extremely meticulous in the care they take in maintaining their personal vehicles. Some vehicle owners are much less meticulous than others. The habits learned with personal vehicles are often brought into the workplace by team members and used on the job. Some team members might rush to the dealership the moment they noticed the oil dipstick was one millimeter short of the "fill" line. Some team members might continue to operate their personal motor vehicle once this vehicle started using oil. Other team members might be willing to operate their vehicles while adding oil often. Some team members likely do not check their oil levels at all!

As pointed out in this analogy, there exists a wide range of acceptability and a certain personal code or integrity that each of us lives by or accepts. The front-line supervisor needs to understand the individual ranges of acceptability and equate this to acceptability at the level of the current company culture. Input from the maintenance department is probably critical in determining the status of this equipment. The use of oil may or may not be critical depending on the area in which the equipment is being used. If the equipment runs primarily outdoors, the use of small amounts of oil may be inconsequential. If the equipment is being used near a chemical process or storage of sensitive materials, any use of oil may be too much and unacceptable.

In the beginning of this discussion, two relatively simple questions were asked about a relatively typical situation that probably plays out numerous times daily in the industrial environment. In answering these two relatively simple questions, numerous other questions arose. After an investigation of the problem is complete, a reasonable judgment should be made. For most cultures this reasonable judgment will be made by a group. This group typically consists of the team member or team members reporting the problem, the area supervisor, and maintenance. Ideally, a reasonable solution is determined. What will that solution be?—it depends. It depends on a lot of factors, some of which we have outlined, as well as numerous other factors that have not been mentioned. Each situation is unique, and there is no simple answer, except—it depends.

Reasonable judgment is a commonly used phrase. What exactly is reasonable judgment? That's easy: reasonable judgment is "common sense." What exactly is "common sense"? One definition of common sense might be having knowledge of things that are intuitive; however, some difficulties can arise when having knowledge of what is intuitive is different from one person and one team member to another. That is where the company culture comes into play, and for safety issues, the company safety culture comes into play. The judgment that is used to determine whether the equipment that is using oil is sidelined or not is a reflection of the company safety culture.

If the decision made does not consider the hazard and potential for imminent danger, then we have a large safety gap. If the decision is, "We have got to finish this job come hell or high water!" then a close examination and upgrade of our practices should be considered. The judgment that management uses in everyday decision making is a big part of our safety culture. Let's look at some other safety-related decision-making processes that are made on a daily basis in industry.

Let's discuss a situation where a team member experiences a cut or laceration while performing work activity. As soon as the team member realizes that he cut himself, he needs to make a decision. The first decision is whether he should report the injury or self-treat. (Most readers are going to say, "Wait a minute!" This worker *must* report this injury! But haven't we all been in situations where the team member chooses *not* to report because he or she thought the injury was just too minor? I know I have.)

If the cut is minor, the team member might choose to self-treat, and only he would likely realize that an injury was sustained; however, if the worker's decision is to report the cut (and I hope the team member decides to report rather than not report any incident), at this point the supervisor, in conjunction with the team member, needs to make a decision. The supervisor needs to determine the extent of the injury. In this situation, there is a variety of information to consider. The supervisor can use:

- His or her own observations of what the cut looks like
- The supervisor's personal experience in cuts (possibly the supervisor is first-aid trained)
- The team member's description of the severity of the injury
- The team member's thoughts as to how he expects the injury to be addressed
- The amount of pain the team member says he is in

If the facility in question has the luxury of an in-house medical professional such as a nurse, possibly obtaining the nurse's opinion would be a logical choice. Especially in recent times, though, fewer workplaces have the luxury of in-house medical staff or plant nurses. Even if you have a nurse on staff, the joint decision might be that the team member would visit the nurse at lunchtime or on the way out. This would indicate that the injury was believed to be minor.

If the facility does not have a nurse or medical professional handy, someone must determine what type of medical attention the injury warrants. If a co-worker or nearby team member is first-aid trained, we should likely get the trained person to look over the injury. If we do not have a first-aid–trained person nearby, our best professional judgment should be used.

Maybe this injury warrants a soap and water wash, application of an antiseptic, and a bandage. Or maybe the cut warrants a stitch and a thorough irrigation. Management needs to groom the people making these important decisions. If the people making the decisions are too tentative, every minor cut could be a trip to the emergency room. When we reach the other extreme, the misdirected supervisor might ask that a team member return to work when the injury warrants a couple of stitches and examination by a licensed health care professional.

When the supervisor has a good working relationship with team members, these judgment calls become a lot easier. Knowing your people and how they react in different situations can help one make the correct decision when trying to decide whether to send someone to the clinic or doctor. Also, having an adverse relationship between team member and supervisor is a recipe for failure. (*Note*: We talk about building relationships in the chapter dedicated to supervisor training.)

Most safety-related situations at work have a similar flow process:

1. A problem arises and is observed (in the two previous cases, the problem was observed by a team member).
2. The team member makes a value judgment and determines the severity of the problem and whether he can fix the problem himself. If the team member can fix the problem, he makes arrangements to do so.
3. After the situation has been rectified, the team member advises the supervisor what transpired at his leisure.
4. If the team member cannot adequately rectify the problem, he notifies the supervisor.
5. The supervisor reviews the team member's judgment, along with his own, and they jointly determine the severity of the problem.
6. The supervisor, jointly with the team member, determines an appropriate action plan.
7. The action plan is followed and adjusted as needed until the issue is resolved.
8. Using professional judgment is the key to success in problem solving, but anytime anyone is expected to use his professional judgment, there is room for, at least, some harsh criticism and certainly the potential for error.

Example: The Little Cut That Became Something Bigger

Let's take another example. A mechanic changing the cable on a crane gets a small cut right above the knuckle on the middle finger of his hand. It bleeds a little, but he is embarrassed to tell the supervisor because the mechanic was not wearing gloves while performing this task. This mechanic knew he was supposed to be wearing gloves, and in fact he was wearing gloves, but he had removed them just minutes before the cable snapped to get more dexterity. When the supervisor toured the work area, the gloves were still lying on the boom of the crane.

The mechanic decides that a bandaid is in order and goes to the lunchroom, where the first-aid kit is kept. While the mechanic is in the process of getting a bandaid, the supervisor shows up and wants to know what is happening. The mechanic tells the supervisor that he has a minor cut and does not offer any detail about how the incident occurred. After a little investigation, the supervisor gets pertinent information about how the incident occurred. With the help of the mechanic, the supervisor determines that the injury is minor.

In this case, the mechanic advised that there was very little blood, and although his finger was a bit numb, his grip was still strong. The supervisor was content that the incident was minor, and work continued. About half an hour later, the supervisor asked the mechanic for an injury update. The mechanic reported that the situation had worsened. The mechanic reported that although his grip was strong, he could not move his finger back to rested position. The supervisor reexamined the injury and sent the mechanic for a medical evaluation.

It turned out that this mechanic had severed a tendon. Although this tendon worked in the opposite direction of the tendon that gives the finger strength, it was still a serious injury. Surgery was performed on the mechanic that involved probing into his hand and finger to retrieve both tendon ends and then reattach them. This mechanic did not work regular duty for more than a month.

This incident received a thorough investigation. As in most accidents, a variety of factors contributed to the situation. For this accident, a variety of improvements were recommended. Some of the main points brought forth during the investigation were as follows:

- An improved work technique to be developed when working on cable

(continued)

- The use of the proper gloves and personal protective equpment (PPE)
- Prompt medical treatment for all injuries
- A variety of others

I do not necessarily agree with all of these recommendations, but they do point to a lack of safety culture. In a work environment with a developed safety culture, this accident would not have happened. A thorough job safety analysis (JSA) would have been in place that would eliminate the cable injury. If this step broke down, the mechanic would have been wearing gloves. If that step broke down, the relationship between the mechanic and the supervisor would have been one where the mechanic would have received quicker medical attention.

As mentioned in the previous example, using sound professional judgment is the key to problem solving, but anytime anyone is expected to use professional judgment, there is room for criticism and error. You can bet that a lot of criticism from a lot of people surrounded this severed tendon situation. The supervisor took most of the heat for not reacting promptly; however, the supervisor made his judgment based on his experience and observations. He trusted the mechanic when the mechanic advised the supervisor that this injury was minor. After all, the cut looked minor, and there was very little blood. Besides the observations, the supervisor had faith in what the worker told him. Even though the supervisor used logic in handling the situation, he was under scrutiny for some time after this incident occurred.

"Machoism"

The type of incident described in the previous example is sometimes misconstrued as a deliberate coverup. I can understand that viewpoint, but I believe that a contributing factor of incidents similar to this one is "machoism". Many working people have this inner pride or strength of character that prohibits them from admitting when they are hurt. They are rough and tough. In the cowboy era, these were the cowboys who were hit by a bullet but only were grazed. If they were hit, it was only a flesh wound that they spit on art wrapped in a bandana, and continued to punch cows, mend fences, or have shoot-outs.

Machoism is not necessarily for men only. If the injured person happened to be a woman, a woman can be equally or even more macho. These team members, men or women, refuse to let pain or injury stop them from doing their work. Legend tells of women who delivered their babies in the potato field and continued to work out the rest of the day, harvesting their fair share of potatoes while carrying their newborn baby with them next to the potato sack.

No one wants to be considered a whiner or a baby. Reporting minor injuries can cause the team member who reports the injury to be called a whiner, baby, or worse. Some supervisors might accuse the team member of reporting the minor injury as the precursor to suing the company or trying to take advantage of the workers' compensation system. Being tough enough to "handle it" is important, isn't it? After all, most fans are in awe when they realize that their favorite players continue to contribute to the game even though they are obviously wincing in pain. So, playing hurt is part of winning athletic teams just as working with minor injuries is part of a winning company culture, right?

The answer to both of these questions is—it depends. Playing with an injury or working with an injury might be okay. On the athletic field, management uses not only the will of the player, but also the judgment of the coach and the medical experts to help determine if playing hurt is acceptable. At work we must use not only the will of the worker, but also the judgment of the supervisor and the medical experts to determine whether the worker can adequately perform his or her job or be working modified duty or not working at all. We will talk about modified-duty programs later.

We must avoid extremism whether the extremism causes us to give exorbitant amounts of attention to a minor injury or noninjury or to totally disregard a serious injury. Both extremes are inappropriate, and finding the middle of the road seems to make the most sense. Again, the supervisor who has a good working relationship with team members has a better chance of determining whether machoism is making it impossible for the worker to adequately report an injury or not. If the relationship between the supervisor and the team members is strained, the potential for sound judgment is lessened.

INSPECTIONS AT THE FIELD SUPERVISOR LEVEL

Management generally expects a lot from team members and generally expects even more from the front-line supervisors or field supervisors. When a difficult decision needs to be made, management expects field supervisors to use their best professional judgment and to adequately address any problems in the field. With the discussions regarding when to sideline equipment and when to send injured workers to the hospital, field supervisors must be involved.

A supervisor's job is one of the most difficult jobs around. The effective supervisor is constantly making decisions based on professional judgment that the company lives and dies by. The effective supervisor must sometimes insist that the worker who believes he has suffered a minor cut must be assessed for stitches. The insistence must be stated so that the worker realizes that the supervisor has the worker's best interest in mind. The supervisor needs to build a good working relationship with all team members. Above all else, the supervisor must provide leadership, especially in safety-related issues.

As far as inspections are concerned, the supervisors must inspect areas for which they are responsible, address team member concerns, and provide leadership in safety. The supervisor provides some redundancy over the adequacy of the work. Instead of referring to the role as redundant, I believe it would be better stated as an extra measure of safety. So the first test that must be passed comes from the judgment of the worker or operator. Once this has been proven, the supervisor can make his or her judgment.

Let's look at an example. Let's say a crane operator inspects the crane and marks the pretrip inspection as everything okay. Later in the day, the department supervisor notices a significant malfunction in the operation of the crane. The supervisor believes that the malfunction might jeopardize safety and notifies the crane operator. The supervisor stops the operation of the crane and asks the crane operator to bring his pretrip inspection form. The form indicates that everything is okay. The crane operator insists that everything is still okay, while the supervisor insists that a significant malfunction is present. A form of this scenario is probably played out many times during the workday. The happy ending to this dilemma comes when the two, as a group, can come to an agreement based on acceptable specifications or control limits.

If this situation is a significant malfunction, then there are probably engineering control limits for acceptable operation of the crane. These limits will likely be in the form of regulatory limits such as OSHA or manufacturer limits. If it is found that we are out of control limits, an immediate adjustment should be made. When we make this adjustment, we need to keep worker importance and worker safety and welfare as the driving force in our decision making.

Let's turn this situation around a bit and look at the scenario where the worker notes a significant malfunction on the crane and will not use it until it has been repaired. The supervisor arrives in the work area and questions why the crane is not working. After checking the condition of the crane, the supervisor decides that the malfunction is not significant and that the crane should continue to operate. This situation also plays out many times during the workday. The happy ending to this dilemma, just as in the previous situation, comes when the two, as a group, can come to an agreement based on acceptable specifications or control limits. But this situation is a bit more tenuous. If the supervisor insists that the work continue without the team member in full agreement, this can be a belittling experience for the team member. If we go back to Chapter 2, remember that team members will respond in accordance with the Hawthorne Effect when they are treated with respect.

There are a variety of ways to handle this situation. One logical approach would be to have another experienced operator look over the equipment and make a judgment. If the other operator believes that the situation is safe, but the original crane operator is still not satisfied, then an experienced technician from maintenance could be summoned to examine the function in question. If this still does not satisfy the crane operator, make arrangements for a representative from the manufacturer to examine the crane and make a judgment. Sometimes an experienced operator can "feel" when a machine is not functioning properly. I have found this to be the case for experienced and well-respected operators even when the supervisor or maintenance person could not easily find a problem.

The judgment of the supervisor is not necessarily the final judgment. Besides maintenance technicians and manufacturers' representatives, other people might get involved in a situation. These other people might include the general foreperson, safety committee member, or safety manager. The outcome of the situation should include a solution that everyone agrees with. All of those involved should be able to keep their professionalism even though a disagreement has been noted.

MANAGER-LEVEL INSPECTIONS

Managers should be touring work areas at an appropriate frequency to eliminate unsafe acts or conditions that the field supervisor and team member did not catch. Typically, managers have more of a "big picture" overview of the work area rather than the hands-on approach of the field supervisor. The manager might find a physical hazard such as a tripping hazard rather than be knowledgeable about the proper operation of a crane. Managers need to have a genuine interest in what goes on in the field, and they need to show this interest to the team members and field supervisors.

A general inspection form that could be used as a template by manager-level team members performing facility inspections is included as Appendix D of this book. The form is general in nature but has components geared toward construction. The actual inspection form to be used would be much closer aligned to the reality of the site and not contain many items that would not be considered applicable. Again, using forms that closely match site conditions is much better than the alternative. Many inspectors take issue with using forms that are full of the term "not applicable" (N/A).

Also, notice that no numerical value is attached to this inspection form. Inspection forms typically do not contain a numerical rating. This is different from the audit. Audits typically have a numerical rating and/or a score or grade attached to them.

MEASURING FIELD COMPLIANCE USING SAFETY AUDITS

Safety audits are another valuable tool to aid in determining field compliance. Audits can also reveal gaps between field activities and safety programs and plans, but the audit attempts to look at compliance from a different level than an inspection. The audit looks at the "big picture" as far as safety is concerned. This big picture will look at general compliance in safety-related areas. These areas can vary, but typical areas that are examined include the following:

- Safety program
- Site-specific health and safety plan
- Site inspections
- Safety incentive programs
- Air-monitoring programs
- Industrial hygiene-related programs
- Accident occurrences, near accidents, and other incident-related paperwork
- Training records/certifications
- Safety meeting minutes

The effective audit will uncover any weakness or noncompliance in the safety-related areas that are being audited. Audits typically get a score, whereas inspections usually do not. Audits are more of a management tool to determine compliance in a general way. Inspections are a tool used by all levels to determine compliance in a specific area (typically and most often physical safety hazards). Management has been baffled when safety inspections continually reveal no deficiencies, yet accident occurrences seem to be prevalent. The use of a comprehensive audit can help eliminate accident occurrences.

Example: A Typical Safety Audit

> The following scenario describes a typical audit: supervisors, safety managers, and others tour work areas and perform an in-depth check of work areas, including spot-checking training records, medical surveillance, licensing, fit testing, knowledge of the site-specific safety plan, and so on, to measure the different parameters of the total safety program.
>
> Notice that this audit does not include the operator, worker, team member, or the person actually performing the physical work. The operators and other team members may be asked for assistance in the completion of certain pieces of the audit, but they would not be considered part of the audit team. The persons actually performing the physical work should be doing inspections, but not audits.
>
> The audits are usually performed routinely. A typical basis for the audit is annually. They can certainly be done more or less than annually, but once per year is a good starting point, especially if you do not currently have an audit program. You will probably find that the type of deficiencies revealed during an audit are different from those noted in a typical inspection. An inspection may find an item like a tripping hazard or a repair that is necessary. An audit may find a deficiency such as the lack of training or certification. Although not always the case, the fix for an inspection item is usually quicker than the fix for an audit item.
>
> Audit results will typically carry with them a numerical score. This score may be related like a grade in school. For instance, above 90 is commendable, between 80 and 89 needs adjustment, 70 to 79 needs substantial adjustment, and below 70 is a failure.
>
> *(continued)*

Let's look at another example. If the audit finds that a tripping hazard needs to be rectified or a quick repair is necessary, this can often be resolved quickly, usually by the person who normally performs an inspection and not an audit. But if we look at the lack of training or certification, this cannot be rectified as quickly or by the person noting the deficiency. Audits will typically cover some items contained in the inspection. One of the deficiencies that can be determined from the audit results is that the inspection program may be lacking. We will talk about how to handle audit findings in the next chapter.

WHAT TRIGGERS A SAFETY AUDIT

There are two main driving forces for a safety audit to take place. An audit might take place just because the cycle has arrived. In other words, if your company's policy is to perform an audit routinely, let's say yearly, and a year has passed, it is likely that a safety audit should occur. The other driving force is an accident occurrence. If a serious accident occurs and more than a few months have passed since the last audit, this might be enough to spark a safety audit. I realize that having an audit sparked by a serious accident goes against the philosophy of any incident or close call being treated as a serious accident. Even though this situation does not seem to follow my philosophy, in many company cultures this is reality. Although this is reality for many companies, this reality should be changed.

Instead of a safety audit taking place after a serious accident, I would suggest a comprehensive accident investigation. The safety audit can take place when its cycle is due rather than because of a serious accident. The accident investigation, if comprehensive enough, should be able to guide the investigators to the underlying factors and recommendations to prevent a recurrence without having to perform a safety audit.

GETTING AN AUDIT PROGRAM STARTED

There are some key items to remember for an effective safety audit program. We need to start with a comprehensive, detailed, and focused audit form. The focus must be on the items that the company holds near and dear. A generic form can be used, but I believe the audit form is much more effective when it is generated for the facility being audited. I have used generic audit forms that are more than 50 pages long, but out of the 50 pages, fewer than 15 pages applied to the facility being audited. So more than two-thirds of the questions asked were not applicable. Some of the people using this form felt that changes needed to be made so that the audit focused more closely on the facility being audited.

When I am tasked to perform a general safety audit on a facility that I have never been to before, I obtain as much background information regarding the facility as possible that is pertinent to safety culture. This general information is always helpful.

Some questions to ask include the following:

- How large in square footage is the facility?
- How many workers are permanently or otherwise stationed at the facility?
- What do employees do at the facility?
- What products are produced?
- What levels of protection are typically used by the team members there?
- Is there a site-specific health and safety plan covering the facility?
- Is there a general safety manual that covers general tasks performed at the facility?
- Is there a site-specific safety recognition program at the facility?
- Is there an in-house inspection program ongoing at the facility?
- Have there been any outside inspections (fire department, health department, OSHA [state or federal], etc.) at the facility? If so, what were the outcomes?
- Have there been any incidents reported at the facility?
- Have there been past audits at the facility? If so, what were the outcomes?
- Who is the local contact person and the person responsible for safety follow-up?

When the auditor arrives at the facility and meets the contact person(s), the auditor should examine accident records before venturing out of the conference room. While examining these accident records, one can get a feel for potential weaknesses in a program. The auditor should be able to answer with confidence a very important first question: "Have all personne incidents and equipment-related incidents been properly reported and investigated?" It is hoped that the auditor will find evidence that incidents have been properly reported and investigated. This will include close calls where no damage occurred and equipment-related incidents.

In most situations, personal injury incidents will typically get reported sooner or later because of the workers' compensation system. An adequate investigation may or may not follow. In the case of an equipment-related incident where there was no injury requiring medical treatment or only first-aid treatment, you may be able to find verbal reports of the incident but no written incident report. After all, with this type of incident, there is no countercheck system because no recordkeeping is required as it is with the workers' compensation system. So, if I were performing an audit, one area that I would be interested in is the reporting of property damage. For example, if a team member drove a forklift into an overhead door and damaged it, would there be a report investigating the incident on file? What happens when property damage occurs? Do these incidents receive an appropriate investigation? Are forklift operators properly trained? Are disciplinary procedures followed when appropriate? It seems like some companies do a very good job in investigating and following up on these types of occurrences, but many do not.

After looking over the accident records, I typically interview local management for their thoughts on room for improvements in the area of safety. Or, if not room for improvements, reported problems. Some of these reported problems may be perceived by certain members of management as being based in other than reality. Your audit may gather information that builds a case for these perceived problems one way or the other.

Once the initial records review has been completed, the auditor should get out of the office and walk through the facility. On this walk-through, the auditor should look for typical safety hazards. I recently audited a facility that consisted of more than 95 percent office workers and a small shop and dock area. This audit revealed a variety of recommendations for improvement from within the shop, dock, and office areas.

The customer contacts at this facility were confident that I would find no deficiencies within the office areas. At the end of the walk-through, they agreed that they were mistaken. Within their office areas were numerous areas where there was insufficient clearance beneath and around sprinkler heads. There were numerous locations where the fire extinguishers were blocked and missing. There were instances where team members were storing a large volume of material on top of structures that were not designed for storage. There were extension cords that were improperly spliced and being used. There were fire alarm pull stations that were blocked and otherwise inaccessible. Some first-aid kits had not been inspected, and some lights within the exit signs were burned out.

Certainly, none of these items constitutes imminent danger in an office setting, but nonetheless my local contact persons were amazed at my results and agreed with them. They told me that they had looked for all of these potential shortcomings before my arrival and believed that the facility was fully compliant. The local contacts told me (and I believe) that they had walked by these areas repeatedly until the hazards that existed blended into the surroundings and went unnoticed.

Although I found no evidence to the contrary at this particular location, I typically want to see for myself if the right tools are available to perform certain jobs. This might include a movable stairway to reach high office storage or a forklift for moving pallets. If there are back injuries occurring because of what appear to be insufficient material handling devices, this should be noted in the audit. As an auditor I always felt that my job was to see things that the local contacts were not looking at or looked at and did not see as a problem.

I also look for facility damage like that mentioned previously, where a forklift damaged a door. So, if the auditor notices damage to a door, the logical question is, "Where is the incident report that goes with the damage to the door?" And "We don't know" is not the response the auditor is hoping to hear. Besides looking for damage to the facility, I look for damage to mobile equipment like forklifts. If the facility uses a fleet of trucks or trailers, the auditor should take a close look at them. If the auditor sees body damage, there should be an incident report or accident report that matches the damage. Plans should be in the works for repair of any damage. But let's get back to the audit example.

Management of this company was proud of its office space and, although there was room for improvement, it appeared that the local management expended adequate time and effort to keep the facility work environment free from serious hazards.

As a starting point, an example general audit form is included in Appendix E of this book. This general form may give some ideas or concepts that can be used as a starting point from which to build and customize an audit form(s). The form contained in Appendix E has been modified and generalized from its original content. This audit form provides a good framework and foundation for building a customized audit. This generalized audit form contains the following 10 main sections:

- General communications
- General recordkeeping
- Training records
- General information
- Inspections

- Medical surveillance
- Fire prevention
- Personal protective equipment
- Material handling and storage
- Tools and equipment

Example: Auditing Apples to Oranges

> Audits can be extremely effective when looking at companies that perform similar functions at multiple locations. Audits can be less effective and possibly antagonistic when the same form is used to audit facilities that have different safety concerns. For instance, I worked at a chemical distribution company that had two distinct operations. One of the operations was warehousing and distribution. This operation would bring in chemicals in bulk in anticipation of customer needs and distribute them as needed to fill an order. They would receive the chemicals by truck and railcar and store them in a warehouse until needed. This operation involved lots of forklifts, a hot room, a clean room, inside general storage, outside general storage, outside covered storage, and more.
>
> The other operation was called repackaging. This operation received bulk chemicals in tank wagons or rail tank cars and either placed the chemicals in a storage tank or immediately drummed them into smaller containers. Whatever was repackaged would be placed into storage and distributed using the distribution operation. Both of these operations were under my supervision and shared the same physical location; therefore, I was tasked by corporate management to perform the standard corporate safety audit on these two different operations. With the diversity between the two divisions, the use of one audit form just did not seem to fit.
>
> In retrospect, this company should have customized the audit form for each operation that would have fit the particular safety concerns of that operation. The corporate management did not want to deal with different audit forms; they wanted consistency. They believed that using one audit form was the fair way to judge one facility against the other. What might have been an alternative would be to offer one type of audit for the repackaging operations and another audit for warehousing/distribution operations. The comparison then would have been closer to the "apples to apples" comparison.

Audit forms typically carry with them a score. This score provides a numerical comparison of the results of one facility to the other. You can imagine that with the competitive nature of human beings, the score received on audits will be hotly contested. When considering even the general audit form offered in Appendix E, an answer to every question may not be forthcoming from every facility audited; however, when the audit is being performed, a certain part of the facility may be shut down or certain information may be unavailable or not applicable. So, it may turn out that not all questions will be answered. In any case, the auditor needs to stay flexible when performing an audit.

During my later years as a safety professional, I was part of a tremendously successful audit program. A customized audit form was specifically designed to fit this company's needs. This form was used at every facility throughout the company. This company had numerous branch locations throughout the country that did similar work. The locations being audited typically included a branch office that would house management, engineering, technicians, administrative personnel, and field staff. Although the field staff might have a desk or office space assigned within a branch office, much of the field staff spent a large percentage of their time performing fieldwork and spent very little time in the branch office.

Usually within the branch office facility and connected to the office space was a warehouse space. The size of the branch offices varied, as did the amount of office and warehouse space. The branch offices were typically leased space. This meant that if the business activity varied upward or downward, the amount of space could be adjusted upward and downward accordingly by renting different-sized facilities. As business cycles rose and declined, and offices and services merged and divested, there were occasions for certain branch locations to move depending on the lease agreements.

The warehouse space was used to store equipment and supplies to support the field activity. The warehouse would typically contain heavy-duty pallet racking, air-monitoring equipment, miscellaneous instrumentation, piping, pumps of various sizes, metallic fittings, tools, safety equipment, PPE, vehicle accessories and attachments, and lots of miscellaneous items.

Most warehouses would have occasion to need both a flammable liquid storage cabinet and a corrosives cabinet. Both corrosives and flammables were needed to perform work activity. The proper storage of both of these classes of chemicals was very important.

At the typical branch locations, the amount of dirty or hazardous work that went on was limited. The branches typically did engineering/office-type work, with a small amount of maintenance going on at certain times in the warehouse. This form suited this type of business well. It concentrated on office issues but still touched on a few shop safety issues; however, if your business does not match this type of business, this form will likely need some adjustment and modification to be most effective. But let me get back to how an audit should be conducted.

After completing the walk through, I typically look through safety-related paperwork, including the following:

- Safety inspections
- Safety inspection follow-ups
- Five-minute toolbox safety meetings
- Departmental safety meetings
- Air-monitoring records
- Job safety analyses
- Safety incentive award programs and their distributed awards
- Training records

At times I would have to go back into the work area to review and confirm observations before making a finding. Each finding results in a numerical score. The audit findings are discussed with local contact persons. The local contacts and auditors should agree about any finding that the auditor believes indicates a need for improvement. It is important to discuss audit findings with the local contacts so that any misunderstandings are cleared up before finalizing the audit findings.

Final audit findings get sent to a committee that reviews the results of the audits. Because this chapter deals with measuring field compliance, we will not go any further with the subject of dealing with audit findings at this time; however, I will pick up this discussion of how to proceed with implementing safety audit findings in a chapter to follow. At this point, if the reader has followed the discussion to this point, you should be able to measure field compliance with the use of safety inspections and safety audits. For the remainder of this chapter, I will discuss some key points of how to ensure that your safety audits are as fair and accurate as possible.

ENSURE THAT THE AUDIT IS "FAIR"

For an audit program to be effective, it should be fair. The comparisons should be as close to "apples to apples" as possible. Continuity among auditors is key. Just let it get around that one auditor interprets a finding differently than another auditor. The auditor with the reputation of giving out the best scores will become very popular very quickly. Put yourself in the place of a management person whose facility was being audited. If you were an operations manager whose bonus depended on the results of the safety audit, wouldn't you hope to receive the auditor who is likely to give out the best score? I'm sure I would. Using one auditor would be a solution.

But sometimes using one auditor is logistically impossible. So, if you have two or more auditors, it is imperative that they communicate with each other. In the beginning and on several occasions, the auditors should meet each other and audit a facility together. They can exchange their thoughts and agree on interpretations and findings. Maybe your facilities are so large and complex that the audits are done by a team, or possibly by several teams. Next let's look at the makeup of the typical audit team.

The audit team should consist of persons who are not only experienced and knowledgeable but also have good communications and interpersonal skills. It is important that the auditors have a high level of integrity and are respected by the people they audit and who will be reviewing the audit results. If you are a smaller organization, your audit team might be one person. Having one person performing all of the safety audits does have some advantages. In this situation, one advantage would be consistency. If you have many locations within your organization, consistency is important. Having one person performing every audit should make this tool very consistent.

The organization that you work for may be so large or have too many locations or be so diverse that not one person would be expected to audit every facility. The situation and logistics could make one-person auditing impractical. If your audit team consists of more than one person, consistency between audits performed by different auditors may be affected.

SUMMARY

Audits and inspections are useful tools in determining field compliance. Inspections should be performed at a variety of levels and at least once, if not more so, during any one shift. Inspections will concentrate on the here and now. Inspections typically focus on items that determine whether a piece of equipment or a production line is safe to operate or should be sidelined. An audit is more focused on "big picture" items.

Sometimes success comes in small steps. Accomplishing field compliance using tools such as an effective inspection and audit program will likely come in small steps. These tools are all part of a developing safety culture. Each step that is made in the development of safety culture is a step in the right direction.

13
Increasing Program Effectiveness

Resting on our laurels is okay when our programs are perfect and there is no progress to be made; however, I am not aware of any facility whose management would agree that it has no room for improvement. There is always a need to improve. No matter how well we perform, there are always ways to make our performance better. This chapter offers ideas for how to make improvements in safety programs that will help develop safety culture.

In the preceding chapters we determined where we were in the past and established the general state of our current safety culture development. Now we need to concentrate on how to get from where we are to where we want to be. This is easier than one might think. As determinations of shortcomings are made and improvement plans are implemented, the culture should be getting stronger all the time. We all know where we want to be as far as safety culture development is concerned: we want to have a zero-accident/incident/loss culture.

Safety-conscious management is interested in having all team members leave work every day, maybe a little tired from putting in a good day's work, but still in possession of all of their fingers, toes, and body parts. Besides ensuring an environment free from physical hazards, we would also ensure that team members' employment would not jeopardize their long-term health. So, simply stated, we want to provide our team members with an environment free from recognized hazards.

This terminology might sound familiar because it has been the basis of safety and health regulations as stated by OSHA. If we can provide this atmosphere, our chances for success in developing our safety culture increase. For most companies, however, it is a difficult task to get cultures to change, even though the changes will be for the ultimate good of the company and the team members in general.

Once we know where we are, we can map the route to where we want to be. This may sound easy, and it can be, but in order for change to have a chance to take hold, a lot of things must happen. Change takes people and the behavior of people. These people must be dedicated to the change and willing to sacrifice in the short term for a long-term gain. Change typically starts with one person and spreads both horizontally and vertically throughout the organization. This one person who starts the change is usually the person in the position of greatest authority. Maybe it is the president of the company, the CEO, the owner, or maybe the chairman of the board.

This person of greatest authority must believe in the need for change and commit the time and necessary resources to implement the change. The speed at which this type of change can take place will vary. As with most things in life, timing is everything. One of the quickest and most successful safety culture improvements that I was part of came during a merger and acquisition phase of a company. Many of the team members within the company that was being bought out were skeptical of the change, especially in the beginning, but six months into the merger things seemed to start to turn around. After a year into the merger, it became obvious that the culture was changing, and 18 months after the merger there was a noticeable improvement in the safety culture that almost everyone noticed and enjoyed.

In retrospect I believe that the safety culture development changed so rapidly because the whole situation was ripe for change. After all, the ownership of the company was changing. There was a new sheriff in town. The policies and procedures were all being merged from one system to another. Culture changes were taking place throughout the company at all levels: benefit packages were changing; purchasing procedures were changing; even the procedure used to make copies was changing. The whole way of business was changing. The timing was perfect for advancement of safety culture.

SAFETY AUDIT REVIEW BOARDS

In the last chapter we mentioned how safety audits can be a useful tool for measuring where we are. Besides expecting consistent and fair audits, we need to remember that the persons reviewing audit results also have a measure of consistency. Typically committees, safety councils, review boards, and so on are instituted to review audit results and discuss plans for improvement. Except for the safety representative and top management representative, other members of the review board will typically have a fixed time frame that they will remain board members. These boards usually have at least three or four members. In this situation it is customary that two of the members are permanent.

The permanent members include a representative of top management and a safety manager. These two figures provide consistency for the group over the long haul. The other members will likely be on the board for two to four years. The exact makeup of these boards varies depending on the setup of your organization. The larger the group of management and the company, the larger the review board group. Some comfortable number should be reached. The review board should include at least four people, but it should not be so large that it becomes ineffective. Again, depending on the setup of the organization and the group dynamic, this number can vary.

The audit review board receives the comprehensive audits and reviews the results. The recommendations and audit results are shared with the review board (which includes upper management) for final review and approval. The involvement of upper management is of the utmost importance.

First and foremost, upper management is showing an interest in safety. The other board members observe firsthand that safety is a core value with upper management and adopt safety as one of their core values. Second, many of the items that evolve from the audit results will need financing. Upper management will have some especially tough decisions to make when lots of money needs to be spent, but the budget has constraints. Having budget constraints is not uncommon.

When budget constraints are considered, running a company is like running a household in many ways. There are usually lots of places where someone can find to spend money and typically less money available for spending than one would like. Long-term plans and prioritization comes into play. The group must determine the highest priorities and work their way through the list until all points are addressed. Sometimes debt instruments come into the mix. The owners, along with top management and the committee, must determine how much cash should be spent or how much debt the company can handle.

The steering or review committee will study the safety audit findings and determine if the recommendations from the audit team are in line with management expectations and vision. Most, if not all, of the audit team's findings are accepted and approved. After this acceptance and approval by the review committee, the review committee will ask the branch or local management to formulate an action plan for completing the "to-do list." The local management puts together the improvement plan and submits it to the review committee. This improvement plan must include target dates and responsible parties.

The branch or local management works on this improvement plan until completion. Local management have the opportunity to work on the "to-do list" almost at their leisure because it is unlikely that any progress or lack of progress would come to the forefront until the next audit. If the audit process is performed yearly, during the next audit it should be relatively easy to discern if progress has been made or was acceptable. If things go according to the plan, one would be able to review the previous findings and determine if the improvement plan was effective.

It is best when an auditor or a particular audit team repeat audits around three times (or three years). This would allow this auditor or this team to have audited this facility three times. During this time frame auditors can not only build a relationship with local management at the branches, but also ensure that progress is being made. Changing auditors for every audit can lead to confusion and a general lack of continuity. Three years would be the maximum time that the format of the audit would remain unchanged. Because the safety culture would have developed substantially over that three-year period, the audit form would probably need to address a different level of issue.

FACILITATING CHANGE THROUGH PROGRAMS

An effective safety audit process can help develop safety culture. Besides an effective safety audit program, other tools can be used to facilitate improvements in safety culture development. New programs can be started for a variety of reasons. Sometimes an industrial accident resulting in a death or serious injury, fire, explosion, or other catastrophic event may drive the push for radical changes. Sometimes the push can come from the government agencies or competition. Sometimes the push might come from worker dissatisfaction or from an inspection from a regulatory agency. Or the push might come from the top manager who realizes that a change is due.

Whatever sparks the need for change will result in management's belief that the institution of a new program will help provide the necessary push for the organization to reach its goal. If the goal is to improve the safety record or become a zero-incidents culture, then management could institute a program or plan to accomplish this goal.

Trying to determine what type of program an organization needs or will be most effective can be an arduous task. When trying to make this important determination, I do not believe there are any easy answers. Making the choice of programs is only the first step. Even after a program is chosen, lots of things must be considered to ensure that the program is successful. This book does not go into the details of how to make the program of choice successful, but will only go as far as choosing potential future programs.

The subject of choosing the programs is addressed soon, but let me first comment about a program that has been popular in the recent past and continues to be popular now. I am referring to the programs called behavior-based safety programs. I am not an authority on behavior-based safety programs, but it appears from what I have read and observed that this type of program may be another useful tool in the toolbox.

My intention is not to endorse or discourage programs using behavior-based safety techniques. If a behavior-based safety program is working for you or your organization, then I believe you have found an effective tool. If your company has tried behavior-based safety programs and not been successful, it might be time to make another assessment. Maybe success could be achieved if a little more time and effort were expended. Maybe a substantial commitment or a commitment that the present management is not willing to support would be required for success. Maybe a definitive cause for the lack of success could not be determined. Perhaps abandoning the program and retrenching is in order.

There are many reasons why effective programs for one company prove to be ineffective for another. As I mentioned earlier, the biggest change in culture that I was a part of took place shortly after a merger/acquisition. The environment was ready for a change. Company cultures were in flux. Although not everyone was looking forward to changing, most believed that a change was inevitable. A variety of different programs were instituted, and the change in culture and safety culture did come. Although the safety culture changed dramatically, a big reason for the success was excellent timing.

Timing is important when planning a program rollout. After all, when you are trying to reach people with a program, and the program facilitator shows up at a facility, it would likely be more effective if the people that the program was intended to reach were in attendance for the presentation. But timing, and program rollout, are subjects for another time.

CHOOSING A SAFETY AUDIT PROGRAM

Let's return to the subject of choosing an effective program. The effective program will cause the anticipated change. This anticipated change, at least in theory, will be a change in the right direction or an improvement. In order to make this important choice regarding the adoption of programs to effect change, we should use a team or group approach.

Support is gained through consensus. I hesitate to use the phrase "steering committee" because of the negativism that is sometimes associated with it. Nonetheless, I believe that "steering committee" is the proper and best phrase and I will continue to use it. The makeup of a steering committee is critical. The committee members should be experienced and respected throughout the organization. A formal education in safety is not a requirement for membership, but the ability to comprehend safety-related problems, leadership capabilities, dedication, and a desire to improve are important criteria.

The steering committee members must be willing to thoroughly analyze findings and make recommendations for improvement. These findings would likely be presented to the committee by the safety department. There will likely be times when the committee and the safety department are not in full agreement. There will likely be times when trends that are analyzed bring forth no real recommendations for improvement.

I believe, and many safety experts agree, that all accidents are preventable. This is a simple statement, and I do believe it is true; however, there are times when the solution to prevention of the accident does not fall within the responsibility of the accident victim. I have been approached many times and continue to be asked about how to prevent vehicular accidents.

A classic vehicle accident case where an audience does not believe that all accidents are preventable contains a scenario similar to the following: A team member is sitting in a vehicle that is located at a toll booth or on the street and another vehicle hits the stopped vehicle from behind. "Was this accident preventable?" My answer is, "Yes, it was preventable." Even though I believe this accident was preventable, I also believe that it was highly unlikely that the person could have prevented this incident even using every defensive driving technique. The person who is most likely to prevent a recurrence of this type of accident was the person in the moving vehicle.

Example: Allegations Abroad

> While working in a foreign country, one of my co-workers drove the company pick-up truck to town during work hours. He was tasked to pick up some pipe and building materials to keep the project moving. It was in the middle of the day, and the team member was proceeding as directed. He legally parked the pick-up truck in front of the hardware store, went inside, and proceeded to shop for the needed items. While in the store for approximately 15 minutes, a drunken resident drove into the legally parked resident and passed out behind the wheel. The local authorities were summoned and made their investigation. The summary of the findings was that my co-worker was at fault, and the drunken resident was completely exonerated.
>
> This incident caused much concern within the company and the safety department. How could my co-worker be considered at fault from this accident? The answer was actually simple. In this country, foreigners and outsiders in general do not have the same rights as local residents and citizens. My co-worker, besides being a foreigner, did not worship the Koran. The local authorities ruled that he had no rights and placed him in custody. In order to stay on good terms with the client, the company had to terminate him from the project. They ended up doing some behind-the-scenes dealings and removing him from the project.
>
> People have asked how this accident could have been prevented. Again, to look toward prevention, we need to focus on the drunken party and not on the co-worker who had legally parked the pick-up truck and was not even operating the vehicle at the time of the accident. This accident investigation was not done objectively, at least by modern-day U.S. standards. I consider this situation to be terribly unfair. If this situation were to be judged by modern-day U.S. standards, it would likely be deemed unfair. But even though the outcome of the incident may not seem fair to you or me, it was considered fair by the local people and local authorities. These local people commonly accepted these standard practices and lived by them.

The steering committee must determine if the information brought to them for consideration is complete, accurate, and comes from an objective viewpoint. Determining these considerations can be a difficult job for any steering committee. It is easy to request that a trend determination be made, but trying to do a good job of this with only in-house information can be impossible. Early on in this book I used the example of back injuries. If your organization has approximately 1,000 team members, one or two expensive back injuries that occur in a short period will likely have a great impact on the company. Yet trying to analyze two back injuries to prevent a similar occurrence may be an effort in futility.

MAKING REASONABLE RECOMMENDATIONS

Let's take an example. An extremely overweight worker with prior back injuries continues to tap into the company's workers' compensation fund every few years because of recurring back problems. Recommendations to prevent recurrence taken from incident investigations and given to the steering committee included the following:

- Encourage the team member to diet.
- Encourage the team member to get on exercise program.
- Encourage the team member to see a doctor.
- Encourage the team member to lose weight.
- Terminate the team member.
- Have the team member use a back brace.
- Have the team member get help when moving heavy objects.
- Have the team member adhere to the company weight-lifting restrictions.
- Force the team member to retire on disability.

Over the years I have seen all of the above recommendations offered as solutions to back injuries. I think that most readers would agree that many of these recommendations should not be included as recommendations to prevent a recurrence. What the steering committee should do is determine whether back injuries are a trend or a concern. If the answer to that question is yes, a closer look is in order. This closer look may require the committee to examine individual accident reports and medical reports to gather more information.

While distributing these reports to members of the committee, an effort should be made to remove the names, dates, and personal information from the handouts before distribution. Management should make an effort to ensure that information that is considered personal and confidential in these reports remains personal and confidential. This information should be used to prevent a recurrence and not for any other purpose.

Keeping one's objectivity can be difficult. Trying to stay objective can be a much more difficult task when the company is a small, tightly knit group and the number of accidents is also small. In this case, most of the committee already knows whose names and personal information were removed from the handout—everybody knows! Even though this may be the case, the committee must make every effort to remain objective and leave personalities out of the decisions.

In my time as a safety professional (which spans more than 20 years), I have not found anyone who tried to hurt themselves on purpose, although I have been around various members of management who believed that certain team members did so, and did so quite often. I have been around numerous team members who have had recurring back problems over their many years of employment. All of these people realized that losing weight would likely result in less pain and less recurring problems. Some of these team members were successful in losing weight, and some of them were not successful, even though they advised me that they had put in the effort, but did not succeed. I have also been around team members who were not overweight but still were affected by recurring back problems. For this type of injury, most of the recommendations mentioned previously might remain intact. The recommendations that fall out in this scenario are those advising people with back injuries to "diet" or "lose weight."

I always find it amazing that someone with or without a weight problem can easily recommend to someone else to diet or lose weight. The overweight people with back problems whom I have been associated with over the years were all aware that they should lose weight, but few were able to successfully do so. I truly believe that the only recommendation that should come out of an accident investigation should be to follow the doctor's advice for exercise, diet, activity, lifting restrictions, and so on.

A few other recommendations seem to be found on many accident reports where a back injury has occurred that are not really inappropriate, but seem to contribute little. Get a back brace or "back and tummy belt" was almost a universal recommendation some years ago. In the recent past, though, back belts seem to have lost some of their popularity. Available information seems to be mixed on the effectiveness of back belts. Some team members think that these belts are great and swear by them. Others really want nothing to do with them and believe that they are a waste of time. I believe that if we keep human nature in mind here, whether we believe that these belts help or not, if we have a team member who believes that they help, and whether wearing such a belt is included as company policy or not, that team member should be allowed and possibly encouraged to wear this type of device.

The recommendation to get help when lifting or moving heavy objects makes good common sense, but some difficulties appear when the definition of "heavy" seems to change with the physical limitations of different team members. Besides a difference in physical limitations among team members, the awkwardness of the move seems to play a large part in the dynamics of a back injury. Besides, how does the team member know how heavy an object is, especially when he or she has not lifted it before? This is a good question, but the equally good answer is usually not too hard to find. What if the team member is expected to work alone?

If the safety culture at your company is well developed, finding out how much objects weigh before moving them is part of the standard operating procedure, part of the way in which you do business, and part of the safety culture. How you go about finding the weights of unknown objects should be discussed at length at the safe lifting/back safety training. What should happen in the organization that is trying to develop its safety culture is that a considerable amount of time and effort will be spent on training.

Let's return to the typical back injury and determine what some reasonable recommendations might be. If we look at some items that are likely to come out of a steering committee's recommendations to eliminate back injuries besides those mentioned earlier, we will likely find recommendations for safe lifting technique training for all team members at appropriate time frames and encouragement in after-work wellness and fitness programs, if deemed appropriate by the treating physician.

If we consider safe lifting technique training for all team members at appropriate time frames, this recommendation would appear to have universal appeal. Unless there are medical reasons for a team member not to have this training, it should be encouraged. This may seem farfetched or even silly to some, but if your top management participates, this training will be more easily accepted. If we consider encouragement in after-work wellness and fitness programs, I believe this recommendation appears to have universal appeal. Unless there are medical reasons for a team member not being able to participate in this program, it should be encouraged. Again, if top management participates in this type of activity, it will likely be more easily accepted.

Let's look at what these two recommendations would cost the company. If we consider a safe lifting program, and just start with a short video and handout, we can probably complete the program in 10 minutes or so. The program will have a nominal cost, and the recordkeeping would likely expand a current team member's job with some extra responsibility. It might be best to have this training combined with some other type of training, either at the beginning or end of another scheduled training course.

If we look at company participation in a health club, the company can usually incur minimal cost if they so choose, or contribute significantly. If a representative from your company approaches the local health club in an effort to increase membership, there may be a couple of positive outcomes from these dealings. Typically, the health club will offer a discount for your company employees just for the company's encouragement to the group. The net cost to the company can be nothing, but each team member who chooses to join can do so while enjoying a group discounted membership. If the company wants to make the cost even more reasonable, it can subsidize the membership as much as it is willing or able.

Promoting and encouraging good health both on and off the job are enhancements to your safety culture. They typically cost little to administer, they are positive things to do, and they pay off in tremendous dividends in worker satisfaction, good health, and the Hawthorne Effect.

The steering committee will gather information pertinent to safety at their company from various sources, analyze the information, and make recommendations for improvement. The types of information that the committee is considering are accidents, accident trend analyses, and comprehensive safety audits. The steering committee's job includes the following tasks:

- Defining problems
- Determining the action to be implemented to solve the problems
- Setting realistic goals and time frames to reach those goals
- Monitoring the progress and making adjustments as necessary

The steering committee will typically not be in charge of implementation of the recommendations but will closely monitor the implementation process. The steering committee may find glitches within the implementation process that need to be addressed. Even after implementation has been reportedly completed, the situation may still be out of control. The committee may need to reassess the implementation and recommend further improvements as necessary.

Even if the implementation process was properly completed, the steering committee may find that the situation is not in control. The steering committee must continue to analyze, make recommendations, and implement improvements as necessary. The improvement process is something that the steering committee will address as a normal order of business.

Written minutes of what transpires during steering committee meetings, including the status of recommendations, implementing improvements, or analyzing and assessing information, will be kept and distributed. Two important facets that must continue to be emphasized are that the committee must continue to communicate to all levels and keep top management actively involved. There will likely be times when top management cannot attend a steering committee. Depending on how your steering committee is set up, the person in the highest authority may not be an active committee member. Keeping this person in highest authority actively involved and informed is necessary throughout the process.

SPECIFIC CASE ANALYSIS

Let's look at the back injury example. If the steering committee analyzes information and determines that back injuries are a problem at your company, the committee should consider the following factors during the analysis:

- Age of team member
- Department that the team member was injured in
- Prior departments that the injured team member worked in
- The specific job that the team member was performing
- The specific act that the team member was doing when the injury occurred
- The history of different departments where the team member previously worked
- The history of job classifications that the team member held
- The history of actual jobs and tasks performed
- Outside interests, hobbies, or other activities that may have affected the injury

Certainly, medical professionals can assist in determining the cause injuries and offer suggestions for avoiding injury in the future. Regarding outside interests or hobbies or other activities that may have affected the injury, let me offer a couple of examples.

I was informed of a team member who believed that mold in the workplace was seriously affecting her health. A variety of consultants were used to take indoor air-quality samples, including mold samples. The results obtained showed that the work area that was sampled contained less mold than the outside air by a substantial margin. This team member insisted that the workplace environment was causing her medical problems. In fact, this team member had been to her own physician, who recently removed a "huge wad" of moldy substance from one of her sinus passages.

After several interviews, it was determined that this team member had a unique hobby—growing exotic mushrooms. This team member spent much of her time away from work in the basement growing a variety of types of mushrooms. This person had failed to inform the physician regarding her hobby. In fact, she did not think this hobby could possibly have anything to do with her health problems. This team member's physician was instrumental in explaining to her that the hobby was a likely cause of her health concerns and made various recommendations for improving her personal health.

Hearing loss is another problem area that seems to surface often. Even though hearing loss can stem from a variety of causes, many states have workers' compensation laws that seem to favor the worker. I have to make the following statement/explanation to various members of management at various times: "These laws are named Workers' Compensation Acts, not Fairness to Employers Acts." Keeping this in mind, I was involved in a case where a team member appeared to be losing his hearing. He had what was known as a standard threshold shift. This team member had been checked several times by medical experts, who could find no other reason for the loss of hearing except for exposure to high levels of noise. This team member's supervisor was extremely diligent in ensuring that the team member wore hearing protection, even in situations that were borderline with noise levels, but the hearing loss continued through the years. The team member's history did not indicate that he was performing work tasks in areas considered to be high noise areas by plant management. Yet, year after year this team member appeared to be losing his hearing.

It was later determined that besides having a hearing loss, this person had a lifestyle that was not conducive to good health or good hearing. In fact, if I were to leave work at the same time this team member left, I could observe this team member making a stop at the local bar. According to his co-workers, this team member had lost his driver's license due to several citations for drunken driving. He had a habit of walking to the bar immediately after work and, while continuing to drink, would ensure that his love for music was satisfied by playing the jukebox as loud as possible while sitting as close as possible to the noise. According to co-workers, his days at home were also filled with loud music and entertainment, including numerous concerts with live bands. It was no wonder that this team member's hearing was being affected.

I was involved in numerous hearing loss cases where only one of the team member's ears was affected by hearing loss. For many of these cases, it appeared that the solution could be found within the lifestyle outside of work rather than within the workplace. When I worked in areas where team members were hunters, I found that many of those people had hearing loss in only one ear, which had probably been caused by hunting with firearms. If you are a hunter, while looking down the barrel of a rifle, depending on whether you are left or right handed, one of your ears will be extremely close to the barrel of the gun and hence the noise of the discharging shell. It was not uncommon to find this type of ailment, especially in more rural areas.

We have discussed specific back injuries in detail at several different times in this chapter. Let me sum up some final thoughts on this subject. Work-related back injuries seem to be a never-ending area of concern with many companies. An L5 laminectomy seemed to be the operation of choice in years past. In recent times there seem to be fewer back operations and more treatment with prescription medicine, work hardening, and concentrated exercise and other medical breakthroughs.

Still, it seems likely that some people just have a "weak" back. They appear to be "prone" to back injury. I had mentioned in earlier chapters that I do not believe that nothing can be done to help the team member who appears to be accident prone. I talked about "Pete" in an earlier chapter. To refresh your memory, Pete was believed by himself and others to be accident prone. In general he would just get hurt a lot. Pete later became focused on safety in the workplace, and his proneness to accidents disappeared. With Pete this proneness to accidents was temporary and was corrected with the right treatment; however, I believe that being prone to back injury is different from general accident proneness.

Being generally prone to having accidents is caused by a general lack of focus or awareness. This condition is more of a mental state and can usually be corrected with the right treatment. Many people with back ailments or other physical ailments never seem to be totally cured. Even after back operations, physical therapy, back braces, and full releases to regular duty, some people still have back pain and trouble with their back. For some this weakness or trouble with their back seems to last their whole lives.

Many of those afflicted may be in positions that involved little lifting. Age, along with past activity or lack of activity, seems to have a relationship to back injuries. Degenerative back disease and arthritic spines seem to be more common as people age. The debilitating effects of a serious back injury, and the amount of time it can keep a team member from operating in the normal mode, is cause for alarm. Certainly the pain, time off, and expense involved with many of these back injuries is enough to cause anyone to be concerned about preventing these injuries. I think that business has a tendency to spend too much time and effort looking for the fix to the back injury cases that, in all likelihood, could not have been easily prevented. After all, how does one prevent arthritic spines and degenerative backs?

This problem seems to closely mimic the problem with hearing loss. In both instances the employer seems to pick up the tab (financially) for the injury when the contributing factors for the injury might be based on non–work-related activities. So, the activities that team members are doing on their personal time can have a devastating effect on a company's bottom line.

Even if there were no other motivation for a company to encourage wellness, safety, and good health on and off the job, I believe it should be encouraged and become part of the company safety culture. I cannot say that general wellness or being fit can stop an arthritic spine, but most physicians seem to encourage physical fitness and activity as part of a generally healthy lifestyle. It appears that many people are saddled with physical ailments that might be more inherited than developmental, so it should be stressed that before starting any exercise program, the team member's personal physician should be consulted.

SUMMARY

There are no simple or best solutions to fix all safety culture problems. The answers to many questions seem to be either "it depends" or "we don't know." Even though the solutions may not be simple or quick, if you have a determined team, the development of safety culture can be accomplished. Safety culture can improve when:

- The need for a change is realized.
- A determination is made about where the culture should be in comparison to where the culture is at this point in time.

- A plan for improvement is mapped out.
- Training, education, and communication are used to realize the goals.

Keeping in mind the Hawthorne Effect during all phases of the safety culture development process is important. Remember that team members typically want to do a good job and be recognized for doing a good job.

14
Training and Grooming Front-Line Supervisors

Success in the development of safety culture can only occur if there is a concerted group effort. This concerted group effort is the team effort that is one of the major themes that has been promoted throughout this book. Failure or setbacks in culture development can occur for many reasons. Any time there is a team effort, the weakness in one team member can hold back the movement of the entire team. If we consider the team like a chain, the chain is only as strong as its weakest link.

The development of a safety culture can fail for many reasons. Often cited and certainly important is the involvement that top management has in the culture. Ultimately, it is top management's job to lead, guide, and be responsible for running the company. "The buck stops here" attitude that Harry Truman made popular while he was president helps us all to realize that the top management executive is ultimately responsible for the success or failure of the company.

Because top management is held responsible for running the company, it is easy to place the blame for failed safety culture with top management; however, changing culture takes time. It is difficult to make serious changes and smoothly implement those changes in a short period. The reasons for failure need to be identified, and a plan for improvement needs to be agreed on and implemented. There may be a variety of legitimate reasons why top management has not been successful in obtaining the development of safety culture that was planned or hoped for. We discuss some of the obstacles to safety culture development later in this chapter.

Changes in safety culture can take a tremendous amount of time and effort. Although the ultimate responsibility lies at the top, culture development can be thwarted even though the top people in an organization are committed to changes and improvements. If the top management is committed, I believe the primary reason for failure lies in the middle of the organization. Not that the top executive involvement is not crucial, because it is, but more often than not, the top manager is either the driving force or realizes soon into the program that safety culture development is important and becomes a true believer early on. At the bottom of the organizational chart lies the team member whose job it is to actually produce the product and get it out of the door. These team members fully realize the need for safety culture and are usually ready to jump on the bandwagon early on, but middle managers and more specifically the front-line supervisors (FLSs) are usually caught in the middle.

There is some logic as to why FLSs feel as if they are caught in the middle. Most companies have to be concerned about the product or services the company produces. After all, companies are basically production oriented. An emphasis on production is certainly understandable. "If we don't get any product moved out the door, none of us will have jobs soon anyway!" We have probably all heard this quote on too many occasions during our careers. For the mature team member, the statement does not need to be repeated. Everyone from the janitor to the president of the company realizes that this is reality. Although the statement is reality, it does not give the whole picture. In fact, it gives only a small picture. In order to have a successful company, you need to have a successful team. The goal of this team needs to be some combination of better, faster, cheaper, and safer.

A rating to determine which of these items is the most important will not be forthcoming because the order of importance does not really matter. What matters is that a workable team is developing and continually improving. Try to envision all of the different departments of a business and the people who are in these departments as members of a big team. Then take these team members and departments and mold this entity into the form of a perfectly round bowling ball. In the bowling ball concept, it does matter which departments have more people or a bigger budget or (potentially) more importance. What matters is that the company team has been molded into an uninterrupted whole (see Figure 14–1).

If any of the departments or team members gets out of alignment or loses sight of the goals of the company, it creates a void or flat spot in the bowling ball. The bigger the flat spot or the void, the more unlikely that the toss of the bowling ball will result in a strike. In fact, as the bowling ball gets more misshapen, the likelihood of throwing a strike

Training and Grooming Front-Line Supervisors 217

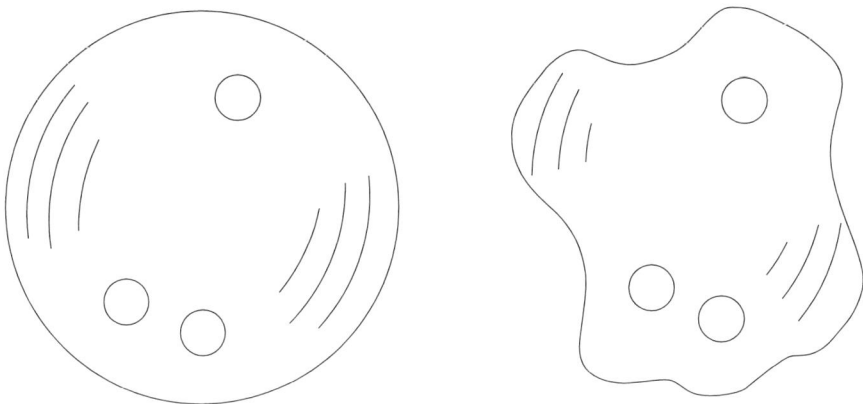

Figure 14–1 Bowling Balls. If all of our company core values could be symbolized by different-colored epoxies that could be rolled up into a perfectly round bowling ball, the final color of the ball would have little to do with rolling a strike. What would be most important would be balance and roundness. When our company values get out of balance, a flat spot is created on the bowling ball. The bigger the flat spot on the bowling ball, the more difficult it is to roll a strike.

decreases while the likelihood of throwing a gutter ball increases substantially.

This bowling ball analogy really hits home. Try to envision all of a company's core values rolled up into a ball. Sure, some people may believe that some departments are more important than others, but just like the Hawthorne Effect, every team member is important. You cannot have a team without every team member.

From a safety viewpoint, common goals and teaming are important concepts. As mentioned previously, the failure of safety culture development can happen for a variety of reasons. Usually these reasons for failure come with no quick-fix or easy answers. Just as in accident investigation, the reasons for failure are usually complicated and multifaceted.

All too often the weak link in the development of safety culture is not the commitment of the CEO or top management person but the strength (or lack of strength) of a company's front-line supervisors (FLSs). In every organization, the FLS usually has one of the most difficult jobs. This chapter concentrates on what can be done to strengthen the skills of the FLS. When FLS skills are enhanced, company culture and in particular safety culture is enhanced. To get a deeper understanding of why FLSs typically need more grooming and training, let's look at how FLSs are chosen.

HOW FLSs ARE CHOSEN

FLSs can arrive at their career path from many different directions. In many instances the FLS will be chosen from the ranks of the team members. This newly chosen FLS may have been chosen because he or she was the most productive member of the department, or maybe the least productive member of the department, a maintenance person who had a knack for keeping the line operating smoothly, the plant manager's brother-in-law, or a whole host of other reasons.

FLSs can be chosen from within the ranks for these reasons or a variety of others. Some FLSs chosen from within the ranks do an excellent job. Unfortunately, some do not. The FLSs chosen from among the ranks have a distinct advantage over those who are hired from the outside because those who come from among the ranks typically know and understand the production process, the people, and the company culture. This knowledge is great and gives the person promoted from within the ranks an advantage over an FLS hired from the outside. If the FLS who was promoted from within the ranks does not have adequate people skills, however, he or she will struggle through his or her career.

Let's look at FLSs who are chosen from outside of the ranks. This newly chosen FLS may have been selected because he or she was the most highly qualified person available after a thorough search, the only person anywhere near qualified who answered the ad, someone who used to work the same job at a competitor, the plant manager's brother-in-law, or a whole host of other reasons. So the reasons that someone becomes a supervisor—whether that FLS has been chosen from within the ranks or hired from the outside—may vary tremendously.

No matter what the reason for becoming an FLS, or why a person was chosen, it will typically be management's goal to ensure that the FLS becomes effective in that position. You can see that it might be difficult for an FLS to be effective in that position if he or she lacks the skills, especially people skills, necessary to function in the FLS position. So, before anyone gets placed in the FLS position, the hiring manager should determine what skills are needed for that spot and hire only those applicants who are the most highly qualified and likely to fit in. Let's look at some of the considerations that the hiring manager should bear in mind for the FLS.

Skill Level, Education, Training, and Experience

Skill level, education, training, and experience are all important considerations for any FLS. The hiring manager should determine what specific skills, if any, are needed for the FLS position that is open. A profile for the skill set of the perfect candidate should be formulated. Maybe an affinity or knack for maintenance, chemical engineering, mechanical engineering, or computer skills will be determined to be important. If we are rating the current FLS, we still need to assess the acceptable skill level and then determine if our FLS is at or above or below these levels of acceptability.

Although we can put together a profile of the skill set that the perfect FLS candidate will possess, the person we choose will probably be far from perfect. Even when we choose the best candidate available, there will probably be areas of the skill set that will be weak or missing. Whatever skills an FLS is lacking, a plan for bolstering this skill level should be shared with the FLS at performance review time. Any plan for skills enhancement should include suggestions for improvement that the FLS can do on his or her own.

Education, training, and experience are all important factors that should be considered when hiring or assessing an FLS. If we are looking to hire an FLS from the outside or promote from within, we need to determine specifics regarding qualifications. We may want to substitute a certain amount of time in service toward educational or other qualifications. No matter what formula we use, we should consider it carefully. After all, these are important career moves that should not be taken lightly.

If we are assessing one of our current FLSs, whatever qualifications this FLS is lacking, a plan for bolstering the skill level should be shared with the FLS at performance review time. Any plan for improving education or training should be included. These plans should also include something that the FLS can bolster on his or her own. Obviously, a lack of experience will take time.

Ability to Accept Change

The FLS must be more willing to accept change than the typical team member. After all, the FLS will be dealing with numerous personalities during the course of the workday. Remaining flexible is essential for successful face-to-face exchanges. FLSs are typically confronted by operational changes and the human factor. Both machines and dealings with humans can offer unique challenges.

Leadership

One of the most important traits that can help make the FLS successful is his or her leadership capabilities. Effective leadership is a tremendous concept, but how does one become a leader? If you look at leadership within our country, there are leaders almost everywhere we look. Let's briefly examine leadership in our government. Typically, we see two types of government officials: those who are elected and those who are appointed. Our president will typically (although not always) start out with a majority of the voters who voted for this leader. During his time in office, the president's popularity (as measured by approval rating polls) can vary tremendously. If the president's approval rating is high, he will probably get reelected. If the approval rating is low, he will probably not be reelected.

A variety of gauges can be used to determine the effectiveness of any leader. One gauge of an effective leader is when that leader is successful in obtaining changes that are unpopular. Sometimes leaders who pushed for an unpopular law find themselves unable to get reelected. Sometimes this ability to pass legislation that is controversial or unpopular more firmly imbeds the leader. Another gauge of an effective leader is when constituents follow this leader during difficult times. When our country struggles through unpopular military conflicts or is going through recession or tightening of the economy, the U.S. president might find that the political party in charge can gain or lose ground depending on the chief executive leadership capabilities.

These leadership gauges can also be used in industry. During the normal course of employment, team members are expected to perform work tasks that they might find less than desirable. Appointed leaders like FLSs do not have to depend on approval ratings or popularity contests; they have a job to perform. If this is the case, it does not really matter what the FLS does because "The FLS is always right and will always be right."

I do not agree. This self-righteous singular attitude may be looked on as belittling and disrespectful of other team members. It would appear that the supervisor who is always right would fall into the theory X manager discussion in Chapter 2. Besides being a theory X manager, if the FLS is always right, the importance of the team seems diminished. If the team members are not important, the Hawthorne Effect will not occur. The FLS should always be doing the "right" thing. Depending on your management philosophy, if you agree with the Hawthorne Effect, most team members are trying to do the right thing. The team members will follow the leader when they are convinced their leader is doing the right thing.

Ethics, trust, and faith are difficult things to measure; however, all of these traits are important to the success of the FLS. We talked about doing the right thing, and the right thing is difficult to determine. Those people judging what the right thing is can be on opposite ends of the spectrum. What is judged as the right thing to one team member could be absolutely the wrong thing to another. When the FLS in known for sincerity, team members will respect and follow the FLS who is trying to do the right thing. Part of this good working relationship that we have talked about throughout the book is based partly on these sometimes elusive values such as ethics, trust, and faith.

Although real-world conditions vary, the ideal situation is when the company philosophy or culture focuses on aspiring to operate with a high ethical level. In previous chapters it has been mentioned that reaching compliance levels is important. Certainly, it is important to reach compliance levels, but it is even more important to realize that reaching compliance levels is only a first step. For the most part, progressive companies need to consider compliance but also institute in-house standards set at levels that are likely to be more stringent than compliance levels to more adequately protect workers, outside interests, and the company. A well-developed safety culture reaches for levels high above regulatory standards. FLSs need to ensure that regulatory standards and in-house standards are met or exceeded. The FLS needs to hold the company values and all ethical considerations in high esteem.

Example: The Case of the Inflated Numbers

> The team members working on step 6 in the widget department turn in inflated output numbers so that they make bonus, safety meeting numbers to make their safety bonus, the number of safety inspections so that they do not get scolded, or any other number in error.
>
> When any numbers get "fudged," there has been an ethical lapse. We need to be encouraged to report the truth. We live by numbers. The numbers that we report need to be accurate. Allowing team members to report things that we know are not accurate cannot be tolerated. After all, if the company we work for came under scrutiny for shady accounting practices and "cooking" the books, this would probably have far-reaching effects on the company's ability to do business. Independent audits makes "cooking" the books sometimes difficult to do, especially for the long term. But sometimes the only person providing a check of numbers reported by the production floor is the FLS. Team members must have trust and faith in the FLS.
>
> *(continued)*

> If the FLS jeopardizes that trust and faith, his or her effectiveness will suffer. When the trust and faith that team members have in the FLS dwindles, their good working relationship erodes along with any thoughts of developing general culture and safety culture.

Mutual Respect

Much of the foregoing discussion is more a matter of philosophy than a matter of easily measured parameters. How can one measure attitude or good working relationships? Attitude surveys have been tried, but their success varies. Also, people's attitudes and approval rating can vary substantially over time. One thing we can hope for is that the color of one's skin, personality conflicts, religious background or beliefs, educational or intellectual levels, and personalities in general should not impede fair treatment.

The FLS will routinely deal with team members from varied backgrounds and levels. The focus must be getting the job done better, faster, cheaper, and safer through team members while remembering the importance of every team member. Treating all team members with respect is a big part of the Hawthorne Effect. People want, need, and should receive respect from everyone, including their FLS. The FLS should ensure that team members are being treated with respect.

When situations arise that cannot be taken care of from within the department, the FLS needs to know who to contact and how to proceed. Safety and the well-being of the team members must be a core value and concern for the FLS.

Ability to Communicate

Throughout this book, communications have been mentioned several times. The ability for an FLS to communicate will probably make a huge difference in his or her success. The novice FLS will quickly develop his or her communication skills. As the career of the novice FLS advances, communication skills grow. Ideally, this skill continues to grow throughout everyone's career and lifetime. Although this development of communications skills will have a positive effect on all of the activities that the FLS becomes involved in, safety culture will also become more developed.

When I began my career as an FLS, one of my mentors stressed the tremendous importance of my attitude. The main point was that I needed to remain positive. Over the years, motivational co-workers, mentors, and members of management have stressed this positivity at times too numerous to mention. I can remember one of my supervisors asking the group, "Is the glass half empty or half full?"

The first time I heard this question, I was not completely sure what the question was supposed to mean. I thought to myself that whether the glass was half empty or half full did not matter—they're the same thing. Later in my career, I had the opportunity to discuss this point with several of my peers. As a scientist, if I held a 1,000-milliliter glass containing 500 milliliters of any liquid, whether I described the glass as half empty or half full it would still contain 500 milliliters. I had difficulty agreeing that it was a more positive statement that a glass was half full rather than half empty. To me half is just half.

On the other hand, perhaps a reflection of attitude can be observed when a group of people go outside at lunchtime and discuss the weather. Are you in the group that says it is partly cloudy or partly sunny? And if you say it is partly sunny, does that make you a more positive person? Perhaps the group that indicates it is partly cloudy is more positive because when it is partly cloudy that must indicate that it is mostly sunny, or does it? I am not sure what it indicates, but being positive is important even though using half empty or half full, or partly cloudy or partly sunny, really does not matter. What does matter is that the FLS tries to remain positive, especially during stressful or trying times.

Balance is an important theme that has been used throughout this book. Remember that extremism in any direction should be avoided. Management needs to stay focused and positive. Team members will probably need to be reminded to stay focused and positive. Although positivity is important, an admission that all is not well is sometimes in order. Being too positive can cause a loss of the respect of fellow team members. On the other hand, let's not be too negative. That same loss of respect can occur and be even more difficult to return to an even keel.

A variety of simple methods can be used to enhance one's communication skills. Try to look at yourself from the vantage point of the way others see you. You may believe that you already practice this method. Just to make sure, have someone videotape you "in the act" of communicating. Have the person with the camera take some closeups and some shots and angles from varied distances. Play close attention to how you look on this videotape. Besides how you look, pay attention to how you sound. After viewing this video, ask yourself some questions: Is the person on the video portraying the person that you believe you are? If not, how can the presentation be adjusted so that you are communicating more effectively? A self-assessment is in order, and a self-improvement plan should be determined and implemented. A couple of points should be emphasized:

- Be yourself.
- Believe in what you say.
- Listen and react to feedback.

While attempting to "be yourself," you will hopefully find that being yourself comes naturally. You might find it difficult to communicate "canned" programs. I had personal difficulty with the glass half empty or half full concept mentioned earlier, and I voiced my opposition to using this phrase as an indication of positivity. Because of my discomfort with this phraseology, I would rarely, if ever, use these terms when communicating with others. Although I did not use this canned phrase or analogy, I would use terms and analogies that I believed in and that came naturally to me.

As one practices communication, a personal style will emerge. Using videos to see how you appear to others around you is a valuable communication improvement tool. But let's say that you do not wish to use a video or are too shy or embarrassed to see and hear yourself on tape. If this is the case, another method can be effective. I locate a mirror, the largest mirror possible, and I practice my communication to myself while observing myself in the mirror. This may seem a bit strange at first, but it helps!

Making an outline or listing key points that you want to communicate can be another useful method in improving communication skills. You might want to communicate a conclusion that you have reached. In order to demonstrate how that conclusion was reached, you might need to share the step-by-step process that was taken in reaching the conclusion. Having these steps written down in outline form beforehand can enhance communication skills and help get the point across effectively. Let's move away from communication improvement skills and back toward real-life industrial situations.

Example: Creative Problem Solving

> Let's say that you are an FLS in charge of step 6 of the widget manufacturing process. Things have not been going too well in the step 6 department lately. A lot of the product that has been produced in the recent past has been out of specification and had to be scrapped and/or recycled. Some of the machinery has shut down completely, so the amount of product being produced has stopped

completely. To make matters worse, there is a rush job from the biggest customer that is being delayed because of the difficulties in step 6.

Things are getting so bad that feed product from step 5 is everywhere! The team members working on steps 7 through 10 are idle, and the team members working on steps 1 through 5 will have to shut down because of space considerations. Everyone seems to be on edge, and the team members in the step 6 department are feeling a lot of heat. Engineering and maintenance seem to be camped out in step 6 department. What should the FLS do?

There are a variety of things that the FLS can do. Since there is downtime, the FLS should make arrangements to fill this downtime with important "fillers." A filler might be a team meeting to discuss important company information. Even better, the filler might be a training session that is overdue or will become due in the near future.

A team meeting might be in order to discuss what has happened and what will be done in the future to prevent the recurrence. It might be advantageous to take the team members out of the department and go to the training room, conference room, lunchroom, or other place to hold your meeting. Sometimes removing team members from the department for a little while can give everyone a fresh perspective toward the problem.

Keep in mind that difficulties similar to the situation described in the previous example will happen. The logic used in accident investigations as a learning experience and a way to avoid a similar situation in the future is a good one to remember for problem solving in general. Typically, the "blame game" serves very little useful purpose. The blame game serves mainly as a finger-pointing experience that pits team members and departments against each other.

For instance, the operations department will typically blame either the manufacturer or maintenance for a poorly designed product or maintenance procedures that were poorly performed or inadequate. Maintenance will typically blame a poor design, "cheap" components, substandard engineering, or others. The engineering department may find fault with operators, often citing "operator error" for failure. Or a variety of departments may blame supervision, the manufacturer, maintenance, or others. Quality control/quality assurance can share in the blame when too "tight" a specification becomes an issue. The human resource department may get blamed for not being able to hire the right people to keep the machinery running. Sales can share in the blame for selling a product that has no profit margin in it, or for promising to deliver the product with too "tight" a production schedule.

The list of who to blame goes on and on. The "blame game" serves very little useful purpose. This blame game should be avoided as much as possible, and the focus should remain on avoidance of future problems.

Human Resources

The effective FLS should have at least a cursory knowledge of the company's human resources policies and procedures. Exactly where does the company stand as far as human relations issues? Knowing when to involve human resources can be difficult to determine. The FLS should not be expected to be a human resources expert, but he or she should know when to get the human resources department involved.

With human resources issues, and almost every other type of issue, establishing and maintaining a good working relationship with all team members will enhance the success of the FLS. Establishing and maintaining a good working relationship is part of an FLS's full-time job. It never stops. Ideally, the relationship between FLSs and other team members will grow and flourish just as the safety culture development grows and flourishes as time passes.

Experienced operators can often tell when a machine is not working properly. They may not be able to determine exactly why the machine is not working properly, but they know that it just does not "feel" right. You might find that you are surprised, as I have been, more often than not, that the experienced operator's judgment proves accurate in the long run, although in the short run the maintenance or engineering team members can literally pull their hair out looking for such a problem. Similarly, as with a machine, the experienced FLS who has established a good working relationship with a team member can tell if something is just not "right" with that team member.

When any team member is just not "right," the FLS needs to realize it and make things happen to change the situation. An exact plan for what buttons to push to make this happen cannot be specified. These types of situations have too many variables to consider to formulate an exact plan. People are so tremendously diverse that no one will have the exact same problem. Even if the problems are similar, there are likely to be numerous actions that yield a workable solution.

Everyone has problems away from work that can affect work behavior. The FLS may feel that outside problems are affecting the team member and feel uncomfortable with any approach. The team member may or may not want to be approached because many people think that nonwork experiences are private. Privacy is certainly an issue that must be respected. If a team member is not interested in discussing nonwork issues, these issues should not be brought up by the FLS. The FLS is not expected to be a psychologist, psychiatrist, or professional counselor; however, when work performance is affected, the FLS should bring this issue to the attention of the team member.

A genuine concern for the team members with whom the FLS works will go a long way in establishing that effective working relationship that we continue to discuss. Even though the FLS is interested in establishing this type of good working relationship, the team member may resist. Sometimes the FLS may bide his or her time but just continue to be positive and focused and accept the relationship. Over time these relationships can grow.

The FLS can remind the troubled team member in a subtle, private way of alternatives that the company has to offer, such as a consultation with the human resources manager or department member. Perhaps the team member needs to be reminded of the Employee Assistance Programs (EAPs) that are typically part of many benefit packages. Keep in mind that the team member and FLS will probably have to work together whether personality conflicts or good working relationships exist or not.

I often equate places of employment to wedding receptions. There is usually a diverse mixture of family and friends. There will be the families of the bride and groom, along with friends and acquaintances of the bride and groom. Everyone shows up to wish the bride and groom well in their future lives together. During the reception, most attendees are on their best behavior and go out of their way to get along with each other. At the end of the reception, all parties go home realizing that they may see many of these same people at family functions, other weddings, funerals, parties, and so on. This best behavior does not have to last too long. Some people leave the reception after dinner. Some people leave after the bar closes. Some people gravitate toward each other and form lifelong happy relationships.

Similarly, at the workplace, some people form tight groups that become close both at the workplace and after work. Some people form close relationships with others but only at work. Yet others seem to have few close friends at work. Although most wedding receptions seem to take place with little personal conflicts, sometimes conflicts can arise. Conflicts can arise even though the parties are trying to be on their best behavior. At a wedding reception, the resolution of conflicts may be easier than at the workplace. After all, if we know a conflict exists, the parties can remain separated or maybe one or both of the parties will not attend the reception.

Conflicts at work may be a lot more difficult to control. Sometimes separating conflicting parties can be difficult to accomplish, especially when one party or the other seems to thrive on the conflict. When these conflicts arise, whether at work or the wedding reception, they need to somehow be resolved. Working these conflicts out can be handled in a variety of ways. The two parties can work together and find a common ground. Sometimes total avoidance is the only way that some people can get along. In principle it should be easy to avoid a conflict at a wedding reception. After all, they usually only last a couple of hours and everyone can gravitate toward those they get along with and hopefully avoid those that they do not get along with for a short period; however, even at wedding receptions, as in life at our places of employment, conflicting parties can often get together and sour the experience. At this point, some type of mediation can be helpful.

This mediation can come from a variety of sources. Sometimes someone in the wedding party might try to act as a liaison or go-between for the two conflicting parties. Rather than mediation, sometimes a distraction or cooling-off period can be effective. In the wedding reception scenario, sometimes a distraction or diversion such as the tapping of spoons on glasses to encourage an embrace between the bride and groom can work well. In a place of employment, there are times when a distraction technique can also help quash conflicts. In the work setting, sometimes other team members can step in and work with the FLS and the team member for a resolution. Sometimes parties get help from outside the work environment that has a positive effect on the work. At times the human resource department might be able to assist with sensitive situations.

Unfortunately, sometimes termination is the outcome, and the road to termination is like the road to divorce. After the reception, sometimes things just go downhill. Some marriages last for many years, and some successful employment situations last for many years. Some marriages end quickly, and some employment situations end quickly. Sometimes the road to termination can be a long and hard road. Sometimes the road to termination is short, with very few twists and turns. Whether the road to termination was long and hard or relatively quick, it is hoped that the parties involved move on to future successes.

Although I believe my marriage has been somewhat successful, I believe that I have also achieved some measure of success in my career path. I attribute much of this success to my ability to stay positive and focused. I continually try to focus on the work task at hand. I personally am trying to improve my work habits, skill level, and productivity as time moves forward. This includes establishing and maintaining good working relationships with my team members while avoiding personality issues and non–work-related distractions.

Avoiding these distractions can be difficult. If you work for a large company but have close long-term relationships with team members both on the job and in personal relationships, avoiding personality issues can be challenging. If you work in a smaller company where team members are tightly knit, this same difficulty can arise. No matter what size organization you work for, keeping a professional outlook and avoiding personality issues should be stressed. We need to remember that the road to termination can be a tough road indeed. Just like the road to divorce, the road to career termination can leave emotional scars.

Novice FLSs learn that a desire to terminate a team member can have numerous negative repercussions. One negative repercussion is that the replacement member is worse than the one who was terminated. Another potential repercussion is that the conflict between parties affects the general culture and safety culture so negatively that a recovery takes an extended period.

THE "GOTCHA" ATTITUDE

The "gotcha" attitude pits management against team members. It is part of a negative spiral: I hurt you, so you hurt me. This downward spiral serves no useful purpose to the participants. As mentioned with the "blame game," the "gotcha" attitude must be avoided. Besides serving no useful purpose to the participants, it can have a negative effect on everyone remotely attached to it and negatively affect general culture and safety culture.

Example: Feuding Factory Workers

> An FLS overheard a conversation regarding illegal drug use by a team member. The FLS notified the human resources manager. Eventually, a drug test was administered, and the drug user was placed in a rehabilitation program and follow-up program. This may appear to be sound practice by the FLS; however, the rest of the details might indicate otherwise.
>
> *(continued)*

The history behind the relationship between the FLS and the drug-using team member turned out to be a "gotcha" relationship. I am not sure how this situation began between these two people, but I am confident that it existed. These relationships can go on so long that the beginnings get cloudy over time. They remind me of the feuds between the Hatfields and the McCoys. To refresh your memory, the feud between these two families went on for generations. The people currently involved in the feuds were not really sure why they were feuding, but they were confident that the feud would continue.

Similar to the Hatfields and the McCoys, the feud between the FLS and the team member continued. Over time the feud became almost common knowledge. The FLS was outspoken and clear that he felt the team member he notified on was a "lowlife." The team member underwent rehabilitation and kept his job. The team member was outspoken that the FLS was picking on him for less than honorable reasons. Every time either one of them had the opportunity to throw mud at the other, the opportunity was taken. Efforts to stop the feud were made. These included separating their work areas, counseling, and so on. I do not believe that the conflict between these two will ever be worked out.

In this example, some basic concepts would have improved the situation. First of all, it appears that the working relationship between the FLS and the team member was not a good one. The word "lowlife" might indicate a lack of respect for the team member. This lack of respect goes against the principles of the Hawthorne Effect. Remember that the importance of every team member and mutual respect is a basic principle that is the foundation for good working relationships. It also appears that the FLS had less than honorable intentions when notifying on the team member. Again, this behavior seemed to point to a lack of respect for the team member.

Confidentiality seemed to be lacking. The FLS was actually proud that he zinged his fellow team member. Many team members expressed feeling that the intention that the FLS had toward the team member was to embarrass him and not to help him. And the FLS let it be known what he did and why he did it in a somewhat public forum. Whether a team member has a drug, medical, alcohol, smoking addiction, or other problem, discussing one team member's problem with other team members is not recommended. Confidential means not to be discussed with other team members who do not need to know that information.

> Besides the need for improvement by the FLS, there is also a need for improvement on the part of the team member. The company at which both of these individuals were employed had a drug use policy. The company screened new hires for drugs and had a yearly (with the physical) and random drug testing policy. The team member who tested positive for drug use was evidently not in compliance with the company policy. When you are doing something that is not in compliance with company policy and get caught, it would seem that the focus should be on fixing the situation (which in this case was EAP, counseling, testing, etc.) and moving on with life.
>
> Looking back on this situation, one can easily find fault and be critical of the situation. Back seat drivers and Monday morning quarterbacks can usually come up with some pretty good solutions to problems after the situation has been resolved. Mistakes are made all the time. The idea is to learn from those mistakes and try to implement changes so that those mistakes are not repeated.

THE SUPERVISOR'S ROLE IN SAFETY INSPECTIONS

What involvement should the FLS have with safety inspections? Ideally, the FLS should be closely involved with safety inspections. The FLS needs to work closely with team members in the area to discuss their safety concerns and include those on the inspection list. Inspection lists should be specifically tuned for conditions in a particular department. If machines are part of the department, they should be included in the inspections. Besides getting input from machine operators, the FLS should get input from a variety of other departments, including maintenance, engineering, QA, and housekeeping.

There are advantages in keeping safety issues as simple and straightforward as possible; however, undoubtedly there will be times when it is necessary to share safety with other core values. Sometimes compromises are made just to keep culture from getting bogged down. Sometimes it may appear that safety issues are being diluted when the programs being instituted focus not only on safety issues but also add quality, production, or other issues whenever possible. Besides what appears to be dilution, there will probably be other situations when the safety and other aspects cannot be separated from each other.

At times it may be best to involve other issues for teaming and other big picture purposes. Again, balance is the key. A judgment needs to be made to determine which items should be included and which should not, or if it makes sense to add another QA checklist or other checklist. Successful safety programs do exist where one inspection was performed that considered a variety of different departmental agendas. It is difficult to say what is needed in any particular situation.

Recently, I was personally involved with a safety reporting system that worked fantastically. Incidents were classified in three groups, and a preliminary report filing period and distribution were set in stone. The supervisors were trained in the procedures. Management dealt quickly with situations where any wavering in the proper procedure was noted. This procedure for incident reporting worked so well that management decided to expand the procedure for a select number of non–safety-related items.

Upper management wanted to ensure that all key people who needed to be informed of certain situations received adequate and prompt notice. The two areas that were added were notices of environmental exceedances or noncompliance and any warrants served. These two parameters were given a numerical classification of an incident and added to the form. Involved parties were retrained in the inclusion of the two new additions on the incident form.

Some members of management felt that the incident reporting system should not be used to report nonsafety items, but upper management would not waver on this subject. Upper management felt that the two nonsafety items that were added were so important that they needed to know about them without fail as soon as possible. The incident reporting system was trusted and found to work. No other system that was devised and used in the company worked as well as the incident reporting system. Because of the reliability of the system and the familiarity that everyone had with the incident reporting system, using it for those two nonsafety items continued with great success.

Maintenance issues are sometimes considered to be more closely related to safety than QA. Getting input from the maintenance department and indications where safety-related problems were found can help the plan eliminate future occurrences. There probably are some items to check that can warn us that something is getting "out of whack." Inspecting for these signs is probably prudent. It could mean that when things are finally "out of whack," an accident or close call may be in the offing.

THE SUPERVISOR'S ROLE IN ACCIDENT INVESTIGATIONS

FLSs should be closely involved in accident investigations. They may or may not be the first ones on the accident scene, but the FLS is probably the most logical choice for knowledge of the team member and the operation surrounding the incident. The establishment of a good working relationship is also important. Accidents are personal and sensitive. The FLS may have to deal with someone in pain or dying. The FLS may have to deal with the loved ones of the accident victim. With a well-developed safety culture, however, it is hoped that the only investigations will be close calls or the review of old accident reports because of an effective zero-incidents culture.

The FLS will probably and should have a large role in any incident investigation. The relationship that has been built between the victim and the FLS should allow for an exchange of pertinent information so that an adequate investigation can be performed. This information should be at least one extensive interview with the accident victim, answering all of the pertinent questions such as who, what, where, how, and why, why, why, and why. A detailed statement of the accident occurrence in the victim's own handwriting should be included. This statement should include the victim's reasoning why this incident occurred and a recommendation from the victim about how to prevent a recurrence. If the victim does not have a good grasp of the English language, an interpreter should be used.

All witnesses and potential witnesses should be interviewed. Witnesses may include not only what team members saw, but also what they heard and when they heard it. If an unsafe act contributed to the incident, you may find no witnesses come forward, even though someone probably *was* an eyewitness. Sometimes a less formal approach will yield more valuable information, especially with an unwilling witness.

When dealing with potentially unwilling witnesses, try to perform all interviewing in as informal a setting as possible. Interview the team member at his or her work station. Hold the interview near the work station rather than in the office, conference room, or other area where the team member may feel intimidated. When fellow team members witness an unsafe act by a co-worker that results in an injury, and that one consequence of management determining that this act was unsafe will probably be the injured employee or other team member getting into trouble, you may find few people willing to discuss what they know.

In some cases a team member may be willing to talk about the accident occurrence, but is not willing to put anything in writing. You may find that the witness does want management to know what happened but does not want to be known as an informant. Depending on the culture, team members may be influenced from two different directions. On the one hand, the team members want management to know what happened and to make the necessary improvements to prevent a recurrence. On the other hand, the team members realize that if they come forward with information, this information may cause a fellow team member embarrassment or even disciplinary action. Many times witnesses seem as if they are caught in the middle. Having the perception of being stuck in the middle is not uncommon.

Any information that helps give a clear picture of an incident occurrence is important. Accurate eye- or ear-witness knowledge regarding accident occurrences should be accepted from any reliable source. This is the case even when the source may not be as cooperative as one would like. If the knowledge is believed to be accurate and useful, it should be used in the investigation even if the information comes without the benefit of a signed statement.

Some investigators use a tape recorder. Unfortunately, the use of a tape recorder or video recorder may yield the investigator mixed results. Certainly, the use of any type of recording device allows an investigator to gather information and record it so that errors and omissions are eliminated during transcription; however, many people are uncomfortable when being interviewed with a tape recorder or video camera. So, the investigator should keep these points in the forefront when choosing circumstances to use a tape recorder or video camera during accident/incident investigation interviews.

At a previous employer's facility, there were no private offices. This facility was cubeland USA; however, there was a conference room for holding meetings. The conference room had only one solid wall. The other three walls were made of solid, clear glass. This conference room was well known for a ventilation system that did not perform too well, too often. Besides poor ventilation, the acoustics in this conference room also seemed to do a poor job of deadening sound. When voices were raised, it appeared as if everyone throughout the office could hear whatever business was going on. Because three of the walls were glass, and it was located in the middle of the office, this room was given the nickname the "fishbowl." There was only one entrance for egress, and everyone in the office who walked by took notice of whoever was sitting in the fishbowl.

As the reader can surmise, the fishbowl was not necessarily the best place to conduct interviews. Besides being considered far from private, it was used for a lot of disciplinary-type discussions, and the open glass and only one door proved intimidating for some. If you had potentially unwilling witnesses, interviewing them in a room like the fishbowl would probably not yield the information you sought.

THE LATE REPORTING OF INCIDENTS

The late reporting of incidents can be a difficult trend to reverse. First of all, no one likes to be the bearer of bad news, and a concerned management team will consider any accident bad news. Second, if your culture is known for accepting late reports without consequences, the trend will probably continue. There probably are a variety of reasons why any particular incident is not reported within the required time frame. A classic excuse still used with some measure of success is that the injured or reporting person was not aware that the incident was a reportable incident until after the acceptable reporting period had lapsed. For personal injuries, maybe an event occurred some weeks ago that was thought to be minor. Maybe the team member reported the injury but did not feel that it warranted incident status. Or maybe the team member failed to report the injury because the injury was believed to be minor at the time of the occurrence.

In the case of property damage incidents, maybe the damage was not noticed until substantial time had passed, and no one admitted to the damage or took responsibility for it. Maybe the damage was not reported because the reporting person is waiting for a repair estimate so that the report can be completed in one step. To some, this option might seem like a good alternative and a savings of time, effort, and paper. It might appear to be good business and common sense. I cannot agree. The reasons for my disagreement will be included a little later on in this chapter, but let's get back to why incidents are reported late.

Another unacceptable excuse used to justify late reporting of accidents/incidents is the lack of understanding by the safety department. For instance, if the safety department does not accept or gives back accident/incident reports that are incomplete and deemed unacceptable, supervisors have been known to use this as an excuse for late reporting. From the supervisor's point of view, there is little reason to turn in a report that is incomplete when that supervisor realizes that the report will be kicked back by the safety department when the shortcoming is realized. In this case, the supervisor might wonder why the report should not just be held until it can be completed properly. Some of the most common unacceptable excuses include the following:

- Information necessary to complete the report is not available.
- All of the facts surrounding the incident are not known.
- Signatories needed to approve the report are on vacation or otherwise not available.
- A copy of the police report will not be available for two weeks.

Late reporting of incidents indicates a need for culture development. Incident reporting should be performed within the required time frames with no room for late reporting. When a team member mentions even the most minor of incidents, the supervisor must realize the potential for this report to change over time. The procedures for reporting such an incident must be adhered to. Even if there is no provision for formally reporting a minor incident, the supervisor should jot down the informal report in his or her planner and note the event for future reference.

There is truly no excuse for late reporting. If a team member reports an injury late for any reason, an appropriate action to alleviate the situation in the future should be implemented. This plan might include counseling, training, noting this performance during the performance review, disciplinary action, or other action.

Very rarely, information will not be readily available to complete an incident investigation. If the investigator is not getting the information needed for an adequate investigation, the investigation may be kept open until such information is available. Rather than delaying the report and withholding information, whatever preliminary information is available should be gathered and distributed through normal channels. A notation should be made that the report being distributed is not yet final. The reasons for the preliminary report and an estimated time for finalization should be offered. In the meantime, if some recommendations can be made for prevention of a recurrence, consider doing so.

Except in the case of vehicular accidents and awaiting police reports, it is rare to find that information is not readily available for incident investigation completion. In most cases, the hold-up is more administrative than substantive. Late reports and investigations should be discouraged. Any incident report should be given a high priority. Being lax with incident reporting is a poor reflection on safety culture. If we feel that accident prevention is a high priority, then accident reporting and follow-up should also be a high priority. Doing these things within the specified guidelines is important. Adherence to deadlines is important in every business. After all, time is money.

Pictures should be taken of the accident scene. If contributory causes of the accident included site conditions, some pictures of those conditions should be taken. People reviewing the report should understand what the conditions were that contributed to the incident. A picture can help recreate the incident scene.

Late reporting should not be tolerated. I believe this is the case even when one is awaiting the development of pictures or vehicular accident police reports. In this day and age, we should consider using a digital camera. Even without the digital camera, one-hour photo developing is available at numerous locations accessible around the country. Make sure that those responsible understand that the accident reports must be published within the time frame called for by the company policy.

If the time frame is not met, this failure should be brought to the attention of a supervisor or human resources staff member for counseling or other action. This might be a consideration at salary review time. Above all, people filling out accident reports must understand the importance of these reports and that timely and complete reporting is something the company cares about. They should also understand that failure to submit a timely report will have consequences.

FLS POSTACCIDENT INVESTIGATION INVOLVEMENT

Besides involvement with accident investigations, the FLS should have a major role in completing the process of getting the injured worker back to work and fully productive. Acclimatization is a principle employed to get team members used to working in temperatures that vary from normal room temperature. This principle "eases" people into working in these extreme temperatures for short periods that increase a little each day or work shift. For full acclimatization, it may take a few weeks. A similar process may be necessary when an injured worker comes back to his or her department to perform work.

Before allowing the injured worker back into the workforce, the FLS should ensure that this team member has been released for duty by the company physician, medical department or personnel, or human resource department. Using the company physician or company department authorization is an important point. Team members have been known to perform a variety of interesting antics both to avoid work and to get back to work.

Example: Getting Official Medical Approval

On one particular remote worksite in the Midwest, one new hire showed up for work without obtaining a medical authorization through the company-approved physician. The FLS turned him away, restating that he would have to have medical clearance before being allowed to start to work. As it turned out, the potential new hire had gone for his physical exam, but the company physician wanted to have a variety of other tests successfully completed before the company physician would deem the potential new hire fit for duty.

The tests that the company physician was requiring were not part of what the company was authorized to perform and pay for during the physical examination; therefore, the team member was told to see his personal physician. The potential new hire's physician would perform the tests (keep in mind that the potential new hire was expected to have these tests performed at his own expense, insurance or otherwise) and then have his personal physician call the company physician. The company physician and personal physician would then consult.

For this particular case, this potential team member had no personal medical insurance and had no personal doctor but had a burning desire to start to work. The potential new hire went to the emergency room of a local hospital, found a doctor, and had some tests performed. The potential new hire then brought a note in from the emergency room doctor clearing this person for work. The potential new hire presented this note to the FLS, who promptly put this newly hired team member to work.

A day or two after the new hire had been allowed to go to work, an in-house medical follow-up person called the FLS and suggested that a problem might exist. The FLS reviewed the medical clearance of the new hire and determined that the doctor had, in fact, cleared this person for work. Unfortunately, the clearance was not obtained from the company doctor. It was obtained from the team member's personal physician (in this case an emergency room physician). There was never any consultation with the company physician regarding the outcome of these tests. In fact, the tests that were taken were abbreviated at the emergency room because the potential new hire was unwilling to pay for the tests indicated by the company physician. As the situation progressed, it was determined that the personal physician and company physician did not agree that this person was fit for duty.

A situation similar to this example repeats itself many times daily in the normal course of business. We have a worker who wants to work, but who is not deemed fit for duty by a company physician. Further testing may result in the clearance for duty, or it may not. One situation we want to avoid is allowing someone to perform work without proper authorization. In the previous example, the team member was very unhappy about being asked to leave the worksite because of a lack of medical clearance. In this team member's logic, he had already proven that he was fit to perform the job. After all, he had been performing it successfully until his less than acceptable fitness for duty was discovered. He expressed with vehemence that there was absolutely no reason not to allow him to continue.

This FLS did not have a lot of experience in these types of matters. This FLS rarely had incidents on his projects and had never dealt with a similar situation in his career. Even though the FLS had the best of intentions, this situation could have been grim. The amount of liability that the company could have undertaken would have been more than they were prepared to handle.

Example: Abiding by the Doctor's Orders

An equipment operator was struck in the head by a flywheel that shattered from a neighboring machine. It was believed that the operator was okay, but after urging from the FLS, the operator was sent for a medical evaluation. The company physician prescribed some medication and told the operator to take the rest of the day off and come back the next morning before the shift change for another look and potential full release. The operator, knowing that his shift was understaffed, went back to work. He told the FLS that he was okay and wanted to get his work completed. The operator went back to the company physician the next day and was released.

In this example, it sounds like things went okay, but I do not agree. I believe that some negatives come into play throughout this situation. If this team member would have hurt himself before receiving a full release form the company physician, this would have been a very unfortunate situation. Our production-oriented supervisor is probably glad to have gotten through the work, but he must realize

(continued)

that getting the work accomplished this way will not be the best for the company in the long run.

These actions cast aspersions on the company physician, medical treatment, and judgment of the company doctor. The order was to go home for the day and be reevaluated tomorrow. The worker chose not to obey the doctor's orders. The FLS needs to ensure that the doctor's orders are being adhered to. The FLS probably committed an error by allowing the team member to return to work without first viewing the return-to-work slip.

This situation does not promote safety culture development. It illustrates a need for improvement in several areas. The relationship between the FLS and the team member needs to be reexamined. It appears as if the good working relationship that has been stressed in this book was lacking. On the surface, it appears that we had a good match. That match is a willing worker and a job that needs to be filled. The worker, besides being willing, really believes that he is physically able to perform the work. The supervisor is more than anxious to get the worker back on board; however, the zeal shared by both the worker and supervisor must be overcome by insisting that the company's requirements be adhered to.

It may be a pleasant experience to play a friendly card game with people whom you trust. Even though the fellow card players trust each other (to a certain extent), the dealer still offers a player the opportunity to cut the cards before the deal. The FLS–team member relationship is similar. Even if the FLS has a good working relationship with a team member, the FLS should check the fitness-for-duty work slip to ensure that the type of work the worker will be tasked to perform falls within any restrictions. Besides checking the fitness-for-duty work slip, a variety of other aspects could have been improved.

Some of the aspects that should have taken place in the previous example are that the FLS or a designate should have accompanied the injured worker to the medical provider and stayed with the injured person until treatment was completed. After that the FLS or designate would have made arrangements to get the injured person home. If this activity were being performed by a designate rather than the FLS, phone calls and communication would be going on so that the FLS knows what the current status is. If arrangements for transportation to the medical provider's establishment need to be made for the follow-up visits, the FLS makes sure that these arrangements are taken care of as well.

This last example appears to be a case of the injured operator being in the wrong place at the wrong time. After all, a flywheel from a neighboring machine flew apart and a piece of this flywheel struck the operator in the head. It is difficult to find fault with this accident being even the partial fault of this operator, but maybe if we change the circumstances a bit we can look at a situation that involves operator error. If we change the example to include certain designated areas for experimental flywheel testing that are supposed to be restricted, shouldn't the operator share some of the blame for the accident? What if the operator was supposed to be wearing head protection but failed to do so? What if the accident was definitely operator error? How does the cause of the accident change the way we treat the injured? What if the FLS is just tired of dealing with silly mistakes that cause accidents?

Finding fault or blame does not build safety culture. Anything that the operator did or did not do that contributed to this accident is certainly important and needs to be considered for ways to prevent the recurrence; however, the injury still occurred, and the first thing we need to do is obtain proper medical treatment for this injury. Whoever or whatever is determined to have caused or contributed to this accident does not play—or at least should not play—into how the accident victim is treated.

At least temporarily, the paperwork surrounding this accident should take a back seat. If the doctor makes recommendations for prescriptions, time off, or restrictions, we should follow those restrictions. If we are unhappy with our doctor's recommendations, then consider finding a new doctor in the future who more closely matches acceptable medical practice. Getting a second opinion, independent medical exam, or other treatment can be done if necessary. There is sometimes a tendency for FLSs to practice medicine. The FLS and all involved should allow the medical staff we have entrusted to do this task to perform their jobs. If we do not trust them, we should seek alternative caregivers.

WORKING WITHIN RESTRICTIONS

If someone is sent for medical treatment for a work-related accident, they will probably return to work with a work restriction. I have seen work restrictions that range from extremely vague to extremely detailed.

An example of a vague return-to-work slip includes: "Have this worker 'take it easy' for a few days. If discomfort persists, return for treatment." I was never quite sure what "take it easy" really meant in medical terms. Should the injured person work at half, three-quarter, one-quarter, or reeeeal slow speed? Maybe I am supposed to get the idea that the physician thinks the injury is minor and will not impede the worker from performing normal tasks at a reduced rate.

On the other hand, some work restrictions contain tremendous detail. An example of what the detailed restriction might say includes: "No bending or stooping. Can lift 40 pounds twice a day, 30 pounds five times per day, 20 pounds twenty times per day, and less than 10 pounds routinely." If I were an FLS and someone returned to work with these restrictions, it would give me the idea that if I did not want to worry too much about staying within these restrictions, I should find a job for this person to perform that requires lifting items that weigh less than 10 pounds, and included no stooping or bending. An FLS must ensure that any job involving modified duty must match the restrictions outlined by the return-to-work slip. This important point needs to be stressed to all FLSs. If someone returns from the medical provider after a work-related injury with a restriction, that person should be assigned only work tasks that fit the work restriction.

As medical treatment and restrictions change, the work tasks must change to fit the restriction. If the FLS does not understand the work restriction or is unclear if a task is okay, consult with the physician. If there is a conflict with what the injured worker believes he or she can do and what the restriction allows, consult with the physician. If the injured worker insists that the restrictions are too "tight," consult with the physician. Effective communication among the FLS, worker, and medical provider must be established and maintained until the injured worker receives a full release.

Working within restrictions is basic to safety culture development. Realizing that the services of those team members are also important should also be kept in mind. Even restricted-duty work tasks should have a purpose and value. Just allowing people to show up when they are on restricted duty is not enough. Remember that following those restrictions is important and cannot be adjusted because of a milestone or safety record.

If the company doctor recommends time away from work because of a serious injury, let's not bring the injured person to work on a stretcher and say that he or she is performing modified-duty work. On the other hand, let's not say that someone cannot contribute and should go home when our physician placed that person on light duty. Again, balance is the key.

Example: Rushing the Healing Process

I was responsible at the corporate level for a project safety manager at a remote construction site in the Midwest. This project safety manager was injured one day on the job. He slipped on some plastic and fell, breaking a bone in his ankle. The subsequent trip to the emergency room resulted in numerous screws installed to keep the bone together, along with a cast. When I first received details of the incident over the phone, I assumed that this accident would be a lost time accident. After all, it seemed from the description that the injured person would probably be in the hospital for several days.

I flew down to the site and met up with some other members of upper management on the way to the site. The three of us had all flown down to help investigate the incident. When we arrived at the site, much to our surprise, we all witnessed the injured employee exiting a vehicle with the help of a co-worker and working his way to his office on crutches. This man was planning on going back to work! None of us could believe it and proceeded to question what was happening. It turns out that the accident occurrence took place earlier the day before. The injured person had surgery later in the day. When the doctor saw him the next morning, he released him to light-duty work, if he could keep his leg elevated for most of the day and perform sitdown work.

All of the different responsibilities and work tasks that this person was supposed to do were discussed. After looking these tasks over, it appeared that this team member could provide a substantial service performing sitdown work at his desk. At this site, one of his jobs was orienting new workers, which took almost two hours and was needed almost daily. Besides the site orientation, this person completed a variety of paperwork on a daily basis.

The next item to be determined was whether this person was truly fit to perform even this type of work. What kind of medication was he supposed to be taking? How would the medication affect his job performance? Was he really going to be able to do something useful, or was it going to be like bringing the injured in on a stretcher and leaning them against a wall? The answers to all of these questions panned out okay. This team member was allowed to continue working on modified duty.

When I think back to this example, I sometimes wonder if this situation added to the development of the safety culture or detracted from it. More than a year later, some team members were discussing

(continued)

this situation and apparently believed that the injured safety person was forced to go back to work. The word on the street about this example was that top executives flew to the site to ensure that this person came back to work promptly. Because I was there, I knew that this was not the case. In fact, the truth lies much closer to the other end of the spectrum, which is that the management was trying, almost desperately, to find a reason to send this man home for at least a few days.

No one wanted to see this site's record blemished, but at the same time, no one wants the modified-duty person to come in on a stretcher. As it turned out, the injured person fought vigorously to be allowed to stay on the job. This injured person was an out-of-towner who would not really be able to do anything but watch TV by himself when he went home anyway. He advised that he was in no pain and was highly motivated to do whatever he could to contribute to his job. He was able to keep his leg elevated perfectly by placing it in the drawer of his desk. Transportation to and from the job for this person inconvenienced no one and did not affect the project in terms of cost or time. The doctor agreed that the injured person could in fact perform work tasks within the restriction and would probably not be affected by the prescription medications except for operating equipment or a vehicle.

Some personal issues seemed to motivate this person more than most injured people. First and foremost, he was a seasoned safety professional with more than 20 years of experience. He did not want the embarrassment of seeing his name affect the perfect safety record. He was determined not to allow this injury to be lost time. The injured person also thought that the accident was in many ways his own fault. Certainly, slippery conditions contributed to this accident, but the accident victim felt that if the same fall occurred to a person who was more agile, younger, in better physical condition, and weighed less, then the broken bone might not have occurred.

We discussed "machoism" in another chapter. This act may have also been influenced by machoism. Maybe he was too proud or too stubborn or too macho to let this accident keep him from working. Maybe he wanted to set an example to other people who might injure themselves and "milk" an injury. I realized that just like the many factors that make up an accident incident, there were many reasons why it was so important for this injury victim to get back to work.

THE SUPERVISOR'S INVOLVEMENT WITH JSAs

If an unsafe act is believed to have been the principal cause of an incident, one needs to raise some questions, such as: Was an adequate JSA performed and available for the task? If so, was the victim aware and familiar with the JSA? If so, what, if anything, did the victim do that was not specified in the JSA? If the answers to these questions indicate a gap, arrangements need to be made to fill the gap. Unsafe acts can and do cause accidents. Eliminating unsafe acts can reduce or eliminate accidents. A thorough JSA is a useful tool to eliminate accidents. If the JSA is not thorough, its effectiveness will be diminished.

The FLS needs to be actively involved in the JSA process, but for many organizations the FLS has little to do with the JSA. This lack of involvement from the FLS in the JSA is an undesirable process. Involvement with JSAs by both the FLS and the team member performing the work is necessary if one expects to use the JSA as an accident prevention tool. Although I realize that the safety person is often used as a focal point for safety for many issues, having a safety person filling out JSAs without upfront active participation from FLSs and team members performing the work should not be an acceptable practice.

Some useful information on JSAs is offered in Appendix A of this book. Basically, we should be able to get input from our FLSs and team members in order to:

- Define the task.
- Break down the task into as many steps as necessary for analysis.
- Determine the hazards involved with each step.
- Determine the protective measures or systems needed to protect the team member against the hazards. *Note:* The protective systems may include engineering systems, administrative controls, or PPE.

Elaborate systems and forms can be developed for JSAs. We have discussed JSAs at several points in this book. For this chapter, we need to emphasize the involvement of the FLS (and team member) with the JSA.

Personal Protective Equipment

The FLS should be aware of PPE that might be made available to team members for protection against hazards. Remember from the previous section that FLSs and team members are involved with developing JSAs.

A component of the JSA will discuss PPE that will protect the team member against known hazards.

A typical shortcoming for PPE is that it is not used. PPE that is not used cannot provide protection. On the other hand, safety people have a tendency to want to place team members in an impervious bubble. The result is overprotection and a tendency to make the work task more difficult to perform. Balance is again the key. Remember that wearing all this protection comes with a downside. If we look at hearing protection, we need to remember that we are blocking out not only noise at levels that we find harmful but also some noise that we might find useful.

If we look at wearing Tyvek suits and respirators, they can do a great job of protecting workers, but they also carry some risk. PPE can be cumbersome and uncomfortable; it can impede one's movements. It can be considered less than stylish, maybe even ugly by some. A respirator can impede peripheral vision, make it more difficult to breathe, and become very, very, warm.

The FLS should provide input regarding both when PPE is needed and where downgrades or adjustments to PPE should be considered. If the FLS is focused, he or she will be able to point out areas of concern that may not be considered by the safety person or team members. One area where the FLS can really make a difference with PPE is ensuring that PPE assigned to workers is sized properly. These days, we have a tremendous differential in the sizes of our team members. For ease of administration, some companies have gotten into a "one size fits all" mode. For certain items, this may be okay.

I worked with a respiratory protection product whose manufacturer boasted that the space-age material was so pliable that their one size would fit every face. I am not going to comment on whether this truly was universal, but up until I stopped using these respirators, I had not been able to find anyone who could not pass a fit test with that brand respirator. Disposable ear plugs seem to be quite successful, and they are PPE that is basically one size fits all. Many items are adjustable; for example, earmuff-type hearing protectors have adjustable arms that vary with the individual's head. In more recent times, many different models of safety glasses have adjustable temple sidepieces so that they can accommodate people with varying sizes of skulls. So, there may be cases where one size fits all, but for many types of commonly used PPE, having varied sizes to fit all of your team members is an excellent idea.

It seems like whether you are on an airplane or wearing work gloves, if you are of an average build, you will probably find both the airplane seat and the work glove more comfortable than someone who is larger than the average build. After all, manufacturing items that fit most team members has some logic attached to it. We could also consider making everything a bit extra large. After all, if it is a Tyvek suit, one can just tuck it in and duct tape it! For a glove, just wear some glove liners. For a large boot, just wear a couple of extra pairs of thick stockings. Usually, if something is large, we can work around it—right? I don't agree. PPE should fit correctly.

FLSs can show themselves as caring about team members when they ensure that they have PPE on hand that actually fits the team members. There are ways to adjust and work around poorly fitting PPE, but why should anyone have to do this? We should *not* have to! If the FLS has workers who are not 80th-percentile male workers, it behooves the FLS to find out what size PPE the other-than-average worker needs and make sure that these items are available. Every non-80th-percentile team member on the crew will appreciate it. When PPE that is the correct size is available for team members' use (i.e., the PPE that was actually specified for the work task), this says a lot about the safety culture. It says that somebody cares about the team members. When team members are cared about and treated with respect and importance, the Hawthorne Effect will occur.

SUMMARY

Supervisors and particularly FLSs are the backbone of any organization. Taking the time to train and groom the FLS is time well spent. All organizations typically have a universal goal. This goal includes being better, faster, cheaper, and safer than the competition. Supervisors come from all walks of life and are at their current spots for a variety of reasons. To perform the job with the most success, an FLS should possess certain characteristics.

Education, training, and experience are all important factors that should be considered when hiring or assessing a FLS. Whether we are looking to hire an FLS from the outside or promote from within, we need to determine specifics regarding qualifications. We may want to substitute a certain amount of time in service toward educational or other qualifications. No matter what formula we use, we should consider it carefully. After all, these are important career moves that should not be taken lightly.

If we are assessing one of our current FLSs, whatever qualifications this FLS is lacking, a plan for bolstering the skill level should be shared with the FLS at performance review time. Any plan for enhancing education or training should be included. These plans should also include something that the FLS can improve on his or her own. Obviously, a lack of experience will take time.

The FLS must be more willing to accept change than the typical team member. After all, the FLS will be dealing with numerous personalities during the course of the workday. Remaining flexible is essential for successful face-to-face exchanges. FLSs are typically confronted by operational changes and the human factor. Both machines and dealings with humans can offer challenges.

Ethics, trust, and faith are difficult things to measure; however, all of these traits are important to the success of the FLS. The FLS should always make the effort to do things "right." Team members will respect the FLS who is trying to do the right thing. Part of this good working relationship between the FLS and team members should be based on values such as ethics, trust, and faith.

Throughout this book, effective communications have been emphasized. The ability for an FLS to communicate will probably make a huge difference in the success of an FLS. The novice FLS will quickly develop his or her communication skills. As the novice FLS advances, communication skills grow. It is hoped that this skill continues to grow throughout everyone's career and lifetime. Although this development of communications skills will have a positive effect throughout the FLS's activities, safety culture development will also be enhanced.

The "blame game" serves very little useful purpose. Supervisors must make every effort to avoid this blame game as much as possible. The focus should remain on avoiding future problems.

Everyone has problems away from work that can affect work behavior. The FLS may feel that outside problems are affecting the team member and feel uncomfortable with any approach. The team member may or may not want to be approached because many people believe that nonwork experiences are private. Privacy is certainly an issue, and this must be respected. If a team member is not interested in discussing nonwork issues, these issues should not be brought up by the FLS. The FLS is not expected to be a psychologist, psychiatrist, or professional counselor; however, when work performance is being affected, the FLS should bring this issue to the attention of the team member.

The "gotcha" atmosphere must be avoided. I get you and throw mud at you, then you get me and throw mud at me. Some people exercise the gotcha attitude in an effort to make themselves look good in the eyes of the boss. Instead of viewing one of the mudslingers looking better than the other, I typically see two muddy people. The gotcha game serves no useful purpose.

A genuine concern for team members with whom the FLS works will go a long way in establishing an effective working relationship. Even though the FLS is interested in establishing this type of good working relationship, the team member may resist. If the FLS truly has respect for team members and demonstrates this respect, a good working relationship will flourish.

15

The Effects of a Tight Economy on Safety Programs and Culture

A tight economy could have a serious effect on a company's safety culture and programs. After all, when everyone is pulling together to control costs, this means the safety department does too—or does it? The answer to this question is more complicated than one might realize. In most cases, there is very little leeway to cut costs for safety. Let's take an example of a safety incentive plan or safety bonus plan.

SAFETY INCENTIVE PLANS

Working safely should be a condition of employment. People are expected to and must work safely. Their future, the future of their co-workers, and the future of the company depend on this. One worker who is "unsafe" can cost the employer many thousands of dollars in damage with one unsafe act and injure or kill him- or herself, fellow workers, or community members.

Understanding that safety is an important factor in running a successful business, it is a core value that management must recognize. When this recognition is made, management will institute programs to ensure that workers are performing their work activity in a safe and efficient manner. Two tools that can be used to encourage safe and efficient work performance are safety incentive plans/programs or safety bonus programs.

Some companies have chosen to engage in what I refer to as "lucrative" safety bonus programs. These lucrative programs can cost many thousands of dollars per worker and are implemented to save employers money by rewarding those who exhibit outstanding safety performance. Although many experts believe that incentive programs have a positive effect on a business, I believe that there is a point at which the additional costs of these programs increases, and the safety performance remains at a standstill or even decreases.

So, one might ask, what is a "good" number, as far as spending money is concerned, to pick for my employee incentive program? I don't have a firm number to recommend, and each company needs to consider its own position carefully before making this determination. If I were really pressured and forced to choose a number, however, I would offer a range. This range for an incentive per person would be $10 to $30 per month.

Remember that this is a range and not a hard number. At the high end of the range, I believe $30 is adequate. Maybe $50 per month fits the budget and suits the business. But, when the program gets a lot higher than $30 per month per person, I believe the program is getting into what I refer to as "lucrative." These lucrative plans can really defeat the purpose. Remember that this $10 to $30 range is only for a well-defined incentive program. This money would not include other money spent on hats, jackets, T-shirts, promotional items, celebrations, and other awards. We will talk about these other items later.

Let's talk about the bottom range—$10 per month per person. When the program calls for an award any lower than $10 per month, it seems as if few people pay much attention to it. Even at $10 per month, many folks just do not seem to take the programs seriously. In order to maximize the effect of a $10 per month payout, the administrators of programs save the award distribution so that the payout is once per quarter. The quarterly payout cuts down on paperwork and boosts the awards to $30 in most cases. This $30 award, even though it still adds up to $10 per month, seems to be more substantial.

Before we go any further, let's first discuss what is included in a safety incentive program. Or, better yet, let's determine what is *not* included in a safety incentive program. Let's not include attendance, tardiness, training, quality, or any other non–safety-related subjects. The core "stuff" includes work-related accidents, incidents, close calls, and so on, along with positive steps taken to prevent accidents. These positive steps included in the core safety "stuff" are items such as tailgate safety talks, safety observer programs, safety training, safety committee meetings, and other related items. We need to remember that we should be concentrating on work-related core safety stuff. I am not saying that quality, productivity, tardiness, attendance, and the like should not be addressed; just that they should not be addressed in the safety incentive plan. Safety incentive plans should just be for safety.

Now, let's talk about those lucrative safety programs. In my experience, I believe that lucrative programs do little to foster safe worker behavior or safety culture development. (Please refer to Chapter 13 for more details on recommended incentive programs.) Lucrative programs that I have been associated with include awards such as one week's pay per month for supervisors whose crews had no lost time accidents. Another part of this lucrative program included hourly workers getting one day's pay per month for their own individual safety record. As I mentioned earlier in the discussion on this subject, when you institute a lucrative incentive program, you may not be getting the best bang for your buck.

A transit company (large-city, mass-transit, union workforce) institutes an incentive plan where workers can routinely be awarded large-screen TVs, vacations, and gift certificates (ranging from many hundreds to thousands of dollars) for not reporting accidents over time. One side effect from safety incentive plans, especially lucrative ones, is nonreporting or underreporting of incidents. From the workers' viewpoint, they would be discouraged from reporting a somewhat minor accident or even a major injury if they knew that they would not be collecting the lucrative payout. On the other hand, perhaps not reporting the minor injury does not allow management to act to prevent the potentially serious injury somewhere down the road. Lucrative incentive programs should be adjusted as necessary.

ELIMINATION OF LUCRATIVE SAFETY PROGRAMS

So, you might ask, "What can be done to eliminate these programs?" I think a decision to eliminate these programs would be in order. For example, if your programs begin at the beginning of the calendar year, you can make the announcement that this will be the last year for this program. This gives everyone adequate notice that this program will be terminated.

People will certainly be upset by the termination of a lucrative safety incentive program. No one likes it when a perk they have enjoyed is taken away. And the more lucrative the perk that is being removed, the more likely it is that opposition will be fierce. Some will probably cry, moan, and gnash their teeth, but if you give adequate notice, or time this event properly, this too shall pass. You need to consider timing carefully. If the profits of your business have been suffering, it might be the exact time for you to change programs to cut costs. The lucrative programs that give less bang for the buck should be eliminated. The amount of complaints that one hears should be lessened if the timing is right or adequate notice is given.

I believe that most of the complaining folks realize that the deal was lucrative, but are just unhappy that something else has been taken away from them. In the long and short term, I do not believe that there will be too much change in accident statistics. Think about it: would you work any safer if you were offered a color TV if you went a year without an accident? Or do you think it is more likely that if you had a minor accident you would probably not report it if you knew doing so would disqualify you from an award?

I believe, and I have observed, that workers will not report or will underreport injuries when reporting injuries disqualifies them, their coworkers, their supervisors, or others from collecting an award. Peer pressure can be substantial in these cases. If departments are lumped into a reporting group, and each department member would receive an award if the department is accident free for a year, how much pressure would be placed on an employee who was injured at work not to report that injury? And how much extra pressure would be placed on that employee if the program were in the last month of the reporting period? This pressure from co-workers can be substantial.

Example: Underreporting Can Be Serious Business

> I have been involved in accident investigations where nonreporting and underreporting went to extremes. I was involved with an investigation where a carpenter sawed off the tips of two fingers with a radial saw and chose to go home and self-treat rather than report the situation as work related.
>
> An office-type worker is bent over performing filing below a cabinet. This person inadvertently left one of the doors of the cabinet open above. When she stood up, she bumped her head on the cabinet door that was above her, jarring her neck. She chose not to report the injury but chose to go home and self-treat. She later obtained medical attention on her own. She underwent treatment and therapy for more than six months.
>
> She finally realized that the injury was serious and that surgery would be required. At this point, she came in and made the report. This employee was in tears when she gave the report to her
>
> *(continued)*

boss. Her reasoning for late reporting was that she thought her co-workers would belittle her for doing something "silly" and that everyone in the department would be disappointed not to receive safety recognition for that year. She felt that she had basically caused this accident by not closing the cabinet door or by not realizing that she had left it open and avoiding the open door.

In both of these cases, the workers resisted reporting their injuries. In both cases the workers, at least initially, choose to treat their own injuries at home instead of seeking qualified medical help. In both cases there is a likelihood that the overall effect of each injury would have been lessened if the injuries had been promptly reported and proper medical care was obtained in a timely manner.

So, let's look at the big picture. Let's discuss what safety incentive plans are supposed to accomplish and how we can modify these plans during times of economic weakness. Very simply stated, safety incentive plans are one more tool in the toolbox that will help eliminate accidents. How do these incentive plans actually work? I believe that when incentive plans are effective, they will raise the workers' awareness so they do not commit unsafe acts. (Unsafe acts cause most work-related accidents.) So, if we have an effective safety incentive program, we are more likely to have our workers commit unsafe acts and more unlikely to have accidents or losses.

That sounds really good, but if we look at the opposing viewpoint, does this mean that if we eliminate the safety incentive program we are more likely to have accidents or losses? I am not sure if this is the case. For one thing, having or not having an incentive program is not the mainstay (or at least should not be the mainstay) of the safety program. The safety incentive program is just another tool in the toolbox that can help develop safety culture. It is a tool to raise awareness but not change someone's lifestyle or make them rich or poor.

SAFETY AUDITS AND SAFETY INSPECTIONS

We discussed both safety audits and safety inspections in several earlier chapters. Both safety inspections and safety audits should be included as part of any company's safety program. Let's quickly review audits versus inspections. An audit is a more formal, more complete, more time consuming process than an inspection. An inspection may be five minutes in duration and may or may not be documented. An inspection is performed often. Often might be once per day before completing a task or it may be several times per day during various stages of task completion.

An audit may take many hours to many weeks and can assess many safety-related facets of the job. These facets may include documentation of training of all persons working at the facility being audited. The audit may require that the files from another location be reviewed so that documentation of training/qualification can be shown. An audit will probably be submitted to members of upper management or ownership for review. Inspections are typically kept locally or reviewed by middle management.

So, now that we have reviewed differences between audits and inspections, how should cost cutting affect safety inspections? Safety inspections are a form of insurance. I do not believe that times will get so tough that we will discontinue general insurance. Furthermore, I do not believe that we can discontinue our inspection program. An inspection program is just part of running any operation. It should not be discontinued under any circumstances.

So it looks as if inspections will stay, even under the toughest of times, but how about safety audits? Safety audits are different from inspections in ways that affect their value and short- and long-term costs. The effect from an inspection can be immediate. If the inspection indicates an immediate and serious problem, this may force an immediate shutdown. Or if the inspection indicates a tripping hazard, this may call for a cleanup, addition of oil dry, removal of the hazard, or other immediate solution. The safety audit may indicate the need for some immediate changes, but more likely it will indicate the need for some more long-term adjustments such as training. Imminent danger situations can be caught by the audit, but typically the audit will not result in a line shutdown like an inspection would.

So, in a way, safety audits are similar to safety inspections. There is much to be gained in performing audits, inspections, and similar types of activities. These activities are what I consider to be cheap insurance. If we are paying our insurance premiums, we should consider retention of safety audit activities; however, due to the longer-term effects that an audit has when compared to an inspection, some adjustments to the audit program would be acceptable.

For instance, let's say the company instituted a ban on all nonessential air travel. This ban on nonessential air travel would be considered temporary with a time frame not to exceed four months. In this situation, there are a couple of alternatives to consider:

- Argue that the safety audits in the program are "essential" and therefore the ban should not affect them.
- Voluntarily delay audits requiring air travel for three months or the duration of the ban.
- Adjust the audit members to eliminate required travel to the places getting audited.
- Try video conferencing, e-mail, conference calls, and so on to use only local people for the audit program.
- Use some combination of the above choices or others.

There are probably other possibilities to help out the company during cost-cutting measures. There should be very little wiggle room in inspection processes, but there probably is a little more wiggle room in the audit program. Because the audit is dealing with longer-term goals, it may be possible to adjust the program in the near term to help with belt tightening.

SAFETY SERVICE AWARDS

Safety service awards are those awards distributed to workers for years of "safe" service. Milestones such as 1 year, 5, 10, 20, and so on of performance without accident are common in the manufacturing industry. These awards are given out along with years-of-service awards; however, the awards for years of service are meant to promote and reward workers for their company loyalty, which is not necessarily connected with safe work activity. I do not believe that lucrative safety service awards are commonplace in industry in the United States. Therefore, I will not be advising that safety service award programs be modified or canceled; however, I believe that these types of awards could be postponed during belt tightening.

Remember, though, that the people who have earned these awards and are probably expecting them should be given written notice that recognizes them for their accomplishments and advises them of an estimate as to when their safety service award will probably be forthcoming. When and where these awards are given out is also important.

Along with rewarding the individual, we want to recognize this individual's accomplishment. We also would like everyone associated with that individual to realize that a milestone has been reached. Recognition is what we are trying to accomplish. Even if we are too poor to give out the award on time due to belt tightening, we should give out written notice signed by a top company official noting the accomplishment. This written notice should be presented in front of a group of people by a person in authority with a statement of congratulations and a discussion of the temporary hard times, along with thoughts of when the actual award will be distributed.

Most team members will relate to the sacrifices that must be made during lean times and will appreciate the recognition even without the reward. The message we are trying to give is, "The company recognizes you for your accomplishments, and when times improve, you will get your just reward. Just stick around and as time passes and the economic situation improves, so will your future and all of our futures." If the company has gone through tough times in the past, what the company did to overcome these tough times might be pointed out.

SAFETY CELEBRATIONS

Safety celebrations are social events that may occur during work hours or after hours to recognize team members for their safety performance. These celebrations may be used to accentuate a milestone of some kind or may be just a monthly or yearly thing. I believe that celebrations should be used to celebrate specific milestones and not just for monthly or annual "things." So, if there is no specific milestone purpose, we could consider throwing that money into the belt-tightening effort. If we are celebrating another year without a lost time accident, on the other hand, this becomes a celebration for a specific milestone accomplishment and should not be touched during cost cutting.

Although I recommend not eliminating specific milestone celebrations, there are some other things one could do to save money. One of these items might include a safety celebration held on the company premises where the top management acts as the food servers to all those attending the celebration. In the events that I have attended, ribs were catered in, but top officials put on aprons and served all those in the food line. After eating, a small amount of time was allocated for a speech and an open house occurred. The hourly folks who were invited to this celebration were thrilled. They felt like the top management truly cared about them and their safety performance, and I believe they did. I believe that most top managers truly care about safety performance but rarely take enough time to show their true commitment. I believe that doing something along the lines of this activity goes a long way to showing everyone that management cares.

As far as saving money on celebrations, I do not recommend that celebrations be postponed and a promise be issued that we will have our celebration in a year or so. I recommend that safety celebrations be continued but on a smaller scale. If we have been serving steak, maybe we could substitute roast beef and chicken. If we have been serving roast beef and chicken, maybe we could consider pizza. Also, as far as a place for having these celebrations, maybe we could consider saving on a hall by using the company premises. The important point is that these celebrations do a lot to promote safety and could be adjusted during tough times but should not be eliminated.

Remember that recognition and management caring seems to have a direct relationship to the Hawthorne study as discussed earlier. The more recognition, attention, or appreciation and respect that management shows toward team members, the more likely it is that our business will thrive. I believe this is true not only with safety considerations, but also with quality, production, and all facets of the business.

SUMMARY

A tight economy or tough times in any business calls for tough decisions. These decisions call for cost cutting on a variety of fronts. Typically, all areas of a business are closely scrutinized for places to save money. While saving money, though, management would like to find places that will not seriously affect the operation of the business. This task is not easy. Managers who are in the position to make these cutbacks typically do so only after much thought has gone into these tough decisions. During these tough economic times, all departments, including the safety department, will probably be examined for opportunities to cut fat, get rid of "dead wood," and reduce costs wherever possible.

The discussion of eliminating lucrative safety programs is an example of what can be done to save costs. No department, including the safety department, should be considered exempt from cost cutting or reevaluation. After all, the safety department budget—just like every other budget—should be examined occasionally to ensure that the budget is on track.

Safety is, or at least should be, one of the "core values" of any company. Even if safety is considered a "core value," there are places to trim the budget. For instance, if your company has what I describe as a lucrative safety incentive program, this program could become a candidate for cost savings; however, if safety is one of the core values and not much fat went into forming the department or the programs, safety will probably be less affected by tightness in the economy.

Appendix A

Job Safety Analysis

INTRODUCTION

Job-related injuries and fatalities occur every day in the workplace. These injuries often occur because employees are not trained in the proper work procedures. One way to prevent workplace injuries is to establish proper work procedures and train all employees in safer and more efficient work methods. Establishing proper work procedures is one of the benefits of conducting a job hazard analysis—carefully studying and recording each step of a job, identifying existing or potential job hazards (both safety and health), and determining the best way to perform the job or to reduce or eliminate these hazards. Improved work methods can reduce costs resulting from employee absenteeism and workers' compensation, and can often lead to increased productivity. This appendix explains what a job hazard analysis is and contains guidelines for conducting your own step-by-step analysis.

It is important to note that the work procedures in this appendix are for illustration only and do not necessarily include all steps, hazards, or protections for similar jobs in industry. In addition, standards issued by the Occupational Safety and Health Administration (OSHA) should be referred to as part of your overall job hazard analysis. There are OSHA standards that apply to most job operations and that emphasize job hazard analysis. Compliance with OSHA standards is mandatory. Employers in any of the 25 states that operate their own OSHA-approved safety and health programs should check with their state agency, which may be enforcing standards that differ somewhat from the federal standards.

Although this appendix is designed for use by forepersons and supervisors, employees also are encouraged to use the information contained in this booklet to analyze their own jobs, be aware of workplace hazards, and report any hazardous conditions to their supervisors.

A job hazard analysis can be performed for all jobs in the workplace, whether the job task is "special" (nonroutine) or routine. Even one-step jobs, such as those in which only a button is pressed, can and perhaps should be analyzed by evaluating surrounding work conditions. To determine which jobs should be analyzed first, review your job injury and illness reports. Obviously, a job hazard analysis should be conducted first for jobs with the highest rates of disabling injuries and illnesses. Also, jobs where "close calls" or "near misses" have occurred should be given priority. Analyses of new jobs and jobs where changes have been made in processes and procedures should follow. Eventually, a job hazard analysis should be conducted and made available to employees for all jobs in the workplace.

Once you have selected a job for analysis, discuss the procedure with the employee performing the job and explain its purpose. Point out that you are studying the job itself, not checking on the employee's job performance. Involve the employee in all phases of the analysis—from reviewing the job steps and procedures to discussing potential hazards and recommended solutions. You also should talk to other workers who have performed the same job.

THE JOB HAZARD ANALYSIS

Before actually beginning the job hazard analysis, take a look at the general conditions under which the job is performed and develop a checklist. The following are some sample questions you might ask:

- Are there materials on the floor that could trip a worker?
- Is lighting adequate?
- Are there any live electrical hazards at the jobsite?
- Are there any chemical, physical, biological, or radiation hazards associated with the job or are any likely to develop?
- Are tools, including hand tools, machines, and equipment, in need of repair?
- Is there excessive noise in the work area, hindering worker communication or causing hearing loss?
- Are job procedures known and are they followed or modified?
- Are emergency exits clearly marked?
- Are trucks or motorized vehicles properly equipped with brakes, overhead guards, backup signals, horns, steering gear, and identification, as necessary?

- Are all employees operating vehicles and equipment properly trained and authorized?
- Are employees wearing proper personal protective equipment for the jobs they are performing?
- Have any employees complained of headaches, breathing problems, dizziness, or strong odors?
- Is ventilation adequate, especially in confined or enclosed spaces?
- Have tests been made for oxygen deficiency and toxic fumes in confined spaces before entry?
- Are work stations and tools designed to prevent back and wrist injuries?
- Are employees trained in the event of a fire, explosion, or toxic gas release?

Naturally, this list is by no means complete because each worksite has its own requirements and environmental conditions. You should add your own questions to the list. You also might take photographs of the workplace, if appropriate, for use in making a more detailed analysis of the work environment.

Breaking Down the Job

Nearly every job can be broken down into job tasks or steps. In the first part of the job hazard analysis, list each step of the job in order of occurrence as you watch the employee performing the job. Be sure to record enough information to describe each job action, but do not make the breakdown too detailed. Later, go over the job steps with the employee.

Identifying Hazards

After you have recorded the job steps, next examine each step to determine the hazards that exist or that might occur. Ask yourself these kinds of questions:

- Is the worker wearing personal protective clothing and equipment, including safety harnesses that are appropriate for the job?
- Are work positions, machinery, pits or holes, and hazardous operations adequately guarded?
- Are lockout procedures used for machinery deactivation during maintenance procedures?

- Is the worker wearing clothing or jewelry that could get caught in the machinery or otherwise cause a hazard?
- Are there fixed objects that may cause injury, such as sharp machine edges?
- Is the flow of work improperly organized? (e.g., Is the worker required to make movements that are too rapid?)
- Can the worker get caught in or between machine parts?
- Can the worker be injured by reaching over moving machinery parts or materials?
- Is the worker at any time in an off-balance position?
- Is the worker positioned to the machine in a way that is potentially dangerous?
- Is the worker required to make movements that could lead to or cause hand or foot injuries, or strain from lifting—the hazards of repetitive motions?
- Can the worker be struck by an object or lean against or strike a machine part or object?
- Can the worker fall from one level to another?
- Can the worker be injured from lifting or pulling objects, or from carrying heavy objects?
- Do environmental hazards—dust, chemicals, radiation, welding rays, heat, or excessive noise—result from the performance of the job?

Repeat the job observation as often as necessary until all hazards have been identified.

Recommending Safe Procedures and Protection

After you have listed each hazard or potential hazard and have reviewed them with the employee performing the job, determine whether the job could be performed in another way to eliminate the hazards, such as combining steps or changing the sequence, or whether safety equipment and precautions are needed to control the hazards. An alternative or additional procedure is to videotape the worker performing his or her job and analyze the job procedures. If safer and better job steps can be used, list each new step, such as describing a new method for disposing of material. List exactly what the worker needs to know to perform the job using a new method. Do not make general statements about the procedure, such as "Be careful." Be as specific as you can in your recommendations.

You may wish to set up a training program using the job hazard analysis to retrain your employees in the new procedures, especially if they are working with highly toxic substances or in hazardous situations. (Some OSHA standards require that formal training programs be established for employees.) If no new procedure can be developed, determine whether any physical changes, such as redesigning equipment, changing tools, adding machine guards, personal protective equipment, or ventilation, will eliminate or reduce the danger. If hazards are still present, try to reduce the necessity for performing the job or the frequency of performing it. Go over the recommendations with all employees performing the job. Their ideas about the hazards and proposed recommendations may be valuable. Be sure that they understand what they are required to do and the reasons for the changes in the job procedures.

Revising the Job Hazard Analysis

A job hazard analysis can do much toward reducing accidents and injuries in the workplace, but it is only effective if it is reviewed and updated periodically. Even if no changes have been made in a job, hazards that were missed in an earlier analysis could be detected. If an illness or injury occurs on a specific job, the job hazard analysis should be reviewed immediately to determine whether changes are needed in the job procedure. In addition, if a "close call" or "near miss" has resulted from an employee's failure to follow job procedures, this should be discussed with all employees performing the job. Any time a job hazard analysis is revised, training in the new job methods, procedures, or protective measures should be provided to all employees affected by the changes. A job hazard analysis also can be used to effectively train new employees on the steps and job hazards.

On July 17, 1990, OSHA issued a proposed rule for the management of hazards associated with processes using highly hazardous chemicals. The agency finalized this rule, called the *Process Safety Management* Standard, on February 24, 1992. In an appendix to the proposed rule, OSHA discussed several methods of process hazard analysis. That discussion, which may be helpful for those doing job hazard analyses, follows:

What if. For relatively uncomplicated processes, review the process from raw materials to product. At each handling or processing step, "what if" questions are formulated and answered, to evaluate the effects of component failures or procedural errors on the process.

Checklist. For more complex processes, the "what if" study can be best organized through the use of a "checklist," and assigning certain aspects of the process to the committee members with the greatest experience or skill in evaluating those aspects. Operator practices and job knowledge are audited in the field, the suitability of equipment and materials of construction is studied, the chemistry of the process and the control systems are reviewed, and the operating and maintenance records are audited. Generally, a checklist evaluation of a process precedes use of the more sophisticated methods described as follows, unless the process has been operated safely for many years and has been subjected to periodic and thorough safety inspections and audits.

What-if/Checklist. The what-if/checklist is a broadly based hazard assessment technique that combines the creative thinking of a selected team of specialists with the methodical focus of a prepared checklist. The result is a comprehensive hazard analysis that is extremely useful in training operating personnel on the hazards of the particular operation. The review team is selected to represent a wide range of disciplines—production, mechanical, technical, and safety. Each person is given a basic information package regarding the operation to be studied. This package typically includes information on hazards of materials, process technology, procedures, equipment design, instrumentation control, incident experience, and previous hazard reviews. A field tour of the operation is also conducted at this time. The review team methodically examines the operation from receipt of raw materials to delivery of the finished product to the customer's site. At each step, the group collectively generates a listing of "what-if" questions regarding the hazards and safety of the operation. When the review team has completed listing its spontaneously generated questions, it systematically goes through a prepared checklist to stimulate additional questions. Subsequently, the review team develops answers for each question. They then work to achieve a consensus of each question and answer. From these answers, a listing of recommendations is developed specifying the need for additional action or study. The recommendations, along with the list of questions and answers, become the key elements of the hazard assessment report.

Hazard and Operability Study (HAZOP). HAZOP is a formally structured method of systematically investigating each element of a system for all of the ways in which important parameters can deviate from the intended design conditions to create hazards and operability problems. The hazard and operability problems are typically determined by a study of the piping and instrument diagrams (or plant model) by a team of personnel who critically analyze the effects of potential problems arising in each pipeline and each vessel of the operation. Pertinent parameters are selected (e.g., flow, temperature, pressure, and time). Then the effect of deviations from design conditions of each parameter is examined. A list of key words (e.g., "more of," "less of," "part of") are selected for use in describing each potential deviation. The system is evaluated as designed and with deviations noted. All causes of failure are identified. Existing safeguards and protection are identified. An assessment is made weighing the consequences, causes, and protection requirements involved.

Failure Mode and Effect Analysis (FMEA). The FMEA is a methodical study of component failures. This review starts with a diagram of the operations and includes all components that could fail and conceivably affect the safety of the operation. Typical examples are instrument transmitters, controllers, valves, pumps, and rotometers. These components are listed on a data tabulation sheet and individually analyzed for the following:

- Potential mode of failure (i.e., open, closed, on, off, leaks)
- Consequence of the failure; effect on other components and effects on whole system
- Hazard class (i.e., high, moderate, low)
- Probability of failure
- Detection methods
- Compensating provision/remarks

Multiple concurrent failures are also included in the analysis. The last step in the analysis is to analyze the data for each component or multiple component failure and develop a series of recommendations appropriate to risk management.

Fault Tree Analysis. A fault tree analysis can be either a qualitative or quantitative model of all the undesirable outcomes, such as a toxic gas release or explosion, which could result from a specific initiating event. It begins with a graphic representation (using logic symbols) of all possible sequences of events that could result in an incident. The resulting diagram looks like a tree with many branches, listing the sequential events (failures) for different independent paths to the top or undesired event. Probabilities (using failure rate data) are assigned to each event and then used to calculate the probability of occurrence of the undesired event. The technique is particularly useful in evaluating the effect of alternative actions on reducing the probability of occurrence of the undesired event.

Employees have the right to complain to their employers, their unions, OSHA, or another government agency about workplace safety and health hazards. Section 11(c) of the *Occupational Safety and Health (OSH) Act of 1970* makes it illegal for employees to be discriminated against for exercising this right and for participating in other job safety and health-related employee activities. These protected activities include:

- Complaining individually or with others directly to management concerning job safety conditions
- Filing formal complaints with government agencies, such as OSHA or state safety and health agencies, fire departments, etc. (An employee's name is kept confidential.)
- Participating in union committees or other workplace committees concerning safety and/or health matters
- Testifying before any panel, agency, or court of law concerning job hazards
- Participating in walk-around inspections
- Filing complaints under Section 11(c) and giving evidence in connection with these complaints

Employees also cannot be punished for refusing a work assignment if they have a reasonable belief that it would put them in real danger of death or serious physical injury, provided that, if possible, they have asked the employer to remove the danger and the employer has refused; and provided that the danger cannot be eliminated quickly enough through normal OSHA enforcement procedures. If an employee is punished or discriminated against in any way for exercising his or her rights under the OSH Act, the employee must report it to OSHA within 30 days. OSHA will investigate and, if the employee has been illegally punished, OSHA will seek appropriate relief for the employee. If necessary, OSHA will go to court to protect the rights of the employee.

SAFETY AND HEALTH PROGRAM MANAGEMENT GUIDELINES

Effective management of worker safety and health protection is a decisive factor in reducing the extent and severity of work-related injuries and illnesses and their related costs. To assist employers and employees in developing effective safety and health programs, OSHA published recommended *Safety and Health Program Management Guidelines* (*Federal Register* 54(18):3908–3916, January 26, 1989). These voluntary guidelines apply to all places of employment covered by OSHA. The guidelines identify four general elements that are critical to the development of a successful safety and health management program:

- Management commitment and employee involvement
- Worksite analysis

- Hazard prevention and control
- Safety and health training

The guidelines recommend specific actions under each of these general elements to achieve an effective safety and health program. A single free copy of the guidelines can be obtained from the U.S. Department of Labor, OSHA Publications, P.O. Box 37535, Washington, DC 20013-7535, by sending a self-addressed mailing label with your request.

SAMPLE JOB HAZARD ANALYSIS

Cleaning Inside Surface of Chemical Tank—Top Manhole Entry

Steps

1. Determine what is in the tank, what process is going on in the tank, and what hazards this can pose.
2. Select and train operators.
3. Set up equipment.

Hazards

    Explosive gas
    Improper oxygen level
    Chemical exposure
- Gas, dust, vapor:
  - irritant
  - toxic
- Liquid:
  - irritant
  - toxic
  - corrosive
  - heated
- Solid:
  - irritant
  - corrosive

    Moving blades/equipment
    Operator with respiratory or heart problem; other physical limitation
    Untrained operator—failure to perform task

Hoses, cord, equipment—tripping hazards
Electrical—voltage too high, exposed conductors
Motors not locked out and tagged
- Establish confined space entry procedures (OSHA standard 1910.146).
- Obtain work permit signed by safety, maintenance, and supervisors.
- Test air by qualified person.
- Ventilate to 19.5% to 23.5% oxygen and less than 10% LEL of any flammable gas. Steaming inside of tank, flushing and draining, then ventilating, as previously described, may be required.
- Provide appropriate respiratory equipment—SCBA or airline respirator.
- Provide protective clothing for head, eyes, body, and feet.
- Provide harness and lifeline. (Reference: OSHA standards: 1910.106, 1910.146, 1926.100, 1926.21(b)(6); NIOSH Doc. #80-406)
- Tanks should be cleaned from outside, if possible.
- Examination by industrial physician for suitability to work.
- Train operators.
- Dry run. (Reference: National Institute for Occupational Safety and Health (NIOSH) Doc. #80-406)
- Arrange hoses, cords, lines, and equipment in orderly fashion, with room to maneuver safely.
- Use ground-fault circuit interrupter.
- Lockout and tag mixing motor, if present.
4. Install ladder in tank.
5. Prepare to enter tank.
6. Place equipment at tank-entry position.
7. Enter tank.
8. Clean tank.
9. Clean up.

Step Hazard

Ladder slipping
Gas or liquid in tank
Trip or fall
Ladder—tripping hazard
Exposure to hazardous atmosphere
Reaction to chemicals, causing mist or expulsion of air contaminant
Handling of equipment, causing injury
- Secure to manhole top or rigid structure.
- Empty tank through existing piping.

- Review emergency procedures.
- Open tank.
- Check jobsite by industrial hygienist or safety professional.
- Install blanks in flanges in piping to tank (isolate tank).
- Test atmosphere in tank by qualified person (long probe).
- Use mechanical-handling equipment.
- Provide guardrails around work positions at tank top.
- Provide personal protective equipment for conditions found. (Reference: NIOSH Doc. #80-406; OSHA CFR 1910.134).
- Provide outside helper to watch, instruct, and guide operator entering tank, with capability to lift operator from tank in emergency.
- Provide protective clothing and equipment for all operators and helpers.
- Provide lighting for tank (Class I, Div. 1).
- Provide exhaust ventilation.
- Provide air supply to interior of tank.
- Frequent monitoring of air in tank.
- Replace operator or provide rest periods.
- Provide means of communication to get help, if needed.
- Provide tow-man standby for any emergency.
- Dry run.
- Use material-handling equipment.

Appendix B

Tools for a Safety and Health Program Assessment

Adapted from:

Construction Safety and Health Outreach Program
U.S. Department of Labor
OSHA Office of Training and Education
May 1996

INTRODUCTION

There are three basic methods for assessing safety and health program effectiveness. This discussion will explain each of them. It also will provide more detailed information on how to use these tools to evaluate each element and subsidiary component of a safety and health program. The three basic methods for assessing safety and health program effectiveness are:

- Checking documentation of activity
- Interviewing employees at all levels for knowledge, awareness, and perceptions
- Reviewing site conditions and, where hazards are found, finding the weaknesses in management systems that allowed the hazards to occur or to be "uncontrolled"

Some elements of the safety and health program are best assessed using one of these methods. Others lend themselves to assessment by two or all three methods.

Documentation

Checking documentation is a standard audit technique. It is particularly useful for understanding whether the tracking of hazards to correction is effective. It can also be used to determine the quality of certain activities, such as self-inspections or routine hazard analysis.

Inspection records can tell the evaluator whether serious hazards are being found, or whether the same hazards are being found repeatedly. If serious hazards are not being found and accidents keep occurring, there may be a need to train inspectors to look for different hazards. If the same hazards are being found repeatedly, the problem may be more complicated. Perhaps the hazards are not being corrected. If so, this would suggest a tracking problem or a problem in accountability for hazard correction.

If certain hazards recur repeatedly after being corrected, someone is not taking responsibility for keeping those hazards under control. Either the responsibility is not clear, or those who are responsible are not being held accountable.

Employee Interviews

Talking to randomly selected employees at all levels will provide a good indication of the quality of employee training and of employee perceptions of the program. If safety and health training is effective, employees will be able to tell you about the hazards they work with and how they protect themselves and others by keeping those hazards controlled. Every employee should also be able to say precisely what he or she is expected to do as part of the program. And all employees should know where to go and the route to follow in an emergency.

Employee perceptions can provide other useful information. An employee's opinion of how easy it is to report a hazard and get a response will tell you a lot about how well the hazard reporting system is working. If employees indicate that the system for enforcing safety and health rules and safe work practices is inconsistent or confusing, you will know that the system needs improvement.

Interviews should not be limited to hourly employees. Much can be learned from talking with first-line supervisors. It is also helpful to query line managers about their understanding of their safety and health responsibilities.

Site Conditions and Root Causes of Hazards

Examining the conditions of the workplace can reveal existing hazards, but it can also provide information about the breakdown of those management systems meant to prevent or control these hazards.

Looking at conditions and practices is a well-established technique for assessing the effectiveness of safety and health programs. For example, let's say that in areas where PPE is required, you see large and understandable signs communicating this requirement and all employees—with no exceptions—wearing equipment properly. You have obtained valuable visual evidence that the PPE program is working.

Another way to obtain information about safety and health program management is through root analysis of observed hazards. This approach to hazards is much like the most sophisticated accident investigation techniques, in which many contributing factors are located and corrected or controlled.

When evaluating each part of a worksite's safety and health program, use one or more of the above methods, as appropriate.

The remainder of this discussion will identify the components found in each element of a quality safety and health program and will describe useful ways to assess these components.

1. ASSESSING THE KEY COMPONENTS OF LEADERSHIP, PARTICIPATION, AND LINE ACCOUNTABILITY

Worksite Policy on Safe and Healthful Working Conditions

Documentation

If there is a written policy, does it clearly declare the priority of worker safety and health over other organizational values, such as production?

Interviews

- When asked, can employees at all levels express the worksite policy on worker safety and health?
- If the policy is written, can hourly employees tell you where they have seen it?

- Can employees at all levels explain the priority of worker safety and health over other organizational values, as the policy intends?

Site Conditions and Root Causes of Hazards

Have injuries occurred because employees at any level did not understand the importance of safety precautions in relation to other organizational values, such as production?

Goal and Objectives for Worker Safety and Health

Documentation

- If there is a written goal for safety and health program, is it updated annually?
- If there are written objectives, such as an annual plan to reach that goal, are they clearly stated?
- If managers and supervisors have written objectives, do these documents include objectives for the safety and health program?

Interviews

- Do managers and supervisors have a clear idea of their objectives for worker safety and health?
- Do hourly employees understand the current objectives of the safety and health program?

Site Conditions and Root Causes of Hazards

Only helpful in a general sense.

Visible Top Management Leadership

Documentation

Are there one or more written programs which involve top-level management in safety and health activities? For example, top management can receive and sign off on inspection reports either after each inspection or in a quarterly summary. These reports can then be posted for employees to see. Top management can provide "open door" times each week or each month for employees to come in to discuss safety and health concerns. Top management can reward the best safety suggestions each month or at other specified intervals.

Interviews

- Can hourly employees describe how management officials are involved in safety and health activities?
- Do hourly employees perceive that managers and supervisors follow safety and health rules and work practices, such as wearing appropriate personal protective equipment?

Site Conditions and Root Causes of Hazards

When employees are found not wearing required personal protective equipment or not following safe work practices, have any of them said that managers or supervisors also did not follow these rules?

Employee Participation

Documentation

- Are there one or more written programs that provide for employee participation in decisions affecting their safety and health?
- Is there documentation of these activities; for example, employee inspection reports, minutes of joint employee-management or employee committee meetings?
- Is there written documentation of any management response to employee safety and health program activities?
- Does the documentation indicate that employee safety and health activities are meaningful and substantive?
- Are there written guarantees of employee protection from harassment resulting from safety and health program involvement?

Interviews

- Are employees aware of ways they can participate in decisions affecting their safety and health?

- Do employees appear to take pride in the achievements of the worksite safety and health program?
- Are employees comfortable answering questions about safety and health programs and conditions at the site?
- Do employees feel they have the support of management for their safety and health activities?

Site Conditions and Root Causes of Hazards

Not applicable

Assignment of Responsibility

Documentation

Are responsibilities written out so that they can be clearly understood?

Interviews

Do employees understand their own responsibilities and those of others?

Site Conditions and Root Causes of Hazards

- Are hazards caused in part because no one was assigned the responsibility to control or prevent them?
- Are hazards allowed to exist in part because someone in management did not have the clear responsibility to hold a lower-level manager or supervisor accountable for carrying out assigned responsibilities?

Adequate Authority and Resources

Documentation

Only generally applicable

Interviews

- Do safety staff members or any other personnel with responsibilities for ensuring safe operation of production equipment have the authority to shut down that equipment or to order maintenance or parts?
- Do employees talk about not being able to get safety or health improvements because of cost?
- Do employees mention the need for more safety or health personnel or expert consultants?

Site Conditions and Root Causes of Hazards

- Do recognized hazards go uncorrected because of lack of authority or resources?
- Do hazards go unrecognized because greater expertise is needed to diagnose them?

Accountability of Managers, Supervisors, and Hourly Employees

Documentation

- Do performance evaluations for all line managers and supervisors include specific criteria relating to safety and health protection?
- Is there documented evidence of employees at all levels being held accountable for safety and health responsibilities, including safe work practices? Is accountability accomplished through either performance evaluations affecting pay and/or promotions or disciplinary actions?

Interviews

- When you ask employees what happens to people who violate safety and health rules or safe work practices, do they indicate that rule breakers are clearly and consistently held accountable?
- Do hourly employees indicate that supervisors and managers genuinely care about meeting safety and health responsibilities?
- When asked what happens when rules are broken, do hourly employees complain that supervisors and managers do not follow rules and never are disciplined for infractions?

Site Conditions and Root Causes of Hazards

- Are hazards occurring because employees, supervisors, and/or managers are not being held accountable for their safety and health responsibilities?
- Are identified hazards not being corrected because those persons assigned the responsibility are not being held accountable?

Evaluation of Contractor Programs

Documentation

- Are there written policies for onsite contractors?
- Are contractor rates and safety and health programs reviewed before selection?
- Do contracts require the contractor to follow site safety and health rules?
- Are there means for removing a contractor who violates the rules?

Interviews

- Do employees describe hazardous conditions created by contract employees?
- Are employees comfortable reporting hazards created by contractors?
- Do contract employees feel they are covered by the same, or the same quality, safety and health program as regular site employees?

Site Conditions and Root Causes of Hazards

- Do areas where contractors are working appear to be in the same condition as areas where regular site employees are working? Better? Worse?
- Does the working relationship between site and contract employees appear cordial?

2. ASSESSING THE KEY COMPONENTS OF WORKSITE ANALYSIS

Comprehensive Surveys, Change Analysis, Routine Hazard Analysis

Documentation

- Are there documents that provide comprehensive analysis of all potential safety and health hazards of the worksite?
- Are there documents that provide both the analysis of potential safety and health hazards for each new facility, equipment, material, or process and the means for eliminating or controlling such hazards?
- Does documentation exist of the step-by-step analysis of the hazards in each part of each job, so that you can clearly discern the evolution of decisions on safe work procedures?
- If complicated processes exist, with a potential for catastrophic impact from an accident but low probability of such accident (as in nuclear power or chemical production), are there documents analyzing the potential hazards in each part of the processes and the means to prevent or control them?
- If there are processes with a potential for catastrophic impact from an accident but low probability of an accident, have analyses such as "fault tree" or "what if?" been documented to ensure enough back-up systems for worker protection in the event of multiple control failure?

Interviews

- Do employees complain that new facilities, equipment, materials, or processes are hazardous?
- Do any employees say they have been involved in job safety analysis or process review and are satisfied with the results?
- Does the safety and health staff indicate ignorance of existing or potential hazards at the worksite?
- Does the occupational nurse/doctor or other health care provider understand the potential occupational diseases and health effects in this worksite?

Site Conditions and Root Causes of Hazards

- Have hazards appeared where no one in management realized there was potential for their development?
- Where workers have faithfully followed job procedures, have accidents or near-misses occurred because of hidden hazards?
- Have hazards been discovered in the design of new facilities, equipment, materials, and processes after use has begun?
- Have accidents or near-misses occurred when two or more failures in the hazard control system occurred at the same time, surprising everyone?

Regular Site Safety and Health Inspections

Documentation

- If inspection reports are written, do they show that inspections are done on a regular basis?
- Do the hazards found indicate good ability to recognize those hazards typical of this industry?
- Are hazards found during inspections tracked to complete correction?
- What is the relationship between hazards uncovered during inspections and those implicated in injuries or illness?

Interviews

Do employees indicate that they see inspections being conducted, and that these inspections appear thorough?

Site Conditions and Root Causes of Hazards

Are the hazards discovered during accident investigations ones that should have been recognized and corrected by the regular inspection process?

Employee Reports of Hazards

Documentation

- Is the system for written reports being used frequently?
- Are valid hazards that have been reported by employees tracked to complete correction?
- Are the responses timely and adequate?

Interviews

- Do employees know whom to contact and what to do if they see something they believe to be hazardous to themselves or coworkers?
- Do employees think that responses to their reports of hazards are timely and adequate?
- Do employees say that sometimes when they report a hazard, they hear nothing further about it?
- Do any employees say that they or other workers are being harassed, officially or otherwise, for reporting hazards?

Site Conditions and Root Causes of Hazards

- Are hazards ever found where employees could reasonably be expected to have previously recognized and reported them?
- When hazards are found, is there evidence that employees had complained repeatedly but to no avail?

Accident and Near-Miss Investigations

Documentation

- Do accident investigation reports show a thorough analysis of causes, rather than a tendency automatically to blame the injured employee?
- Are near-misses (property damage or close calls) investigated using the same techniques as accident investigations?
- Are hazards that are identified as contributing to accidents or near-misses tracked to correction?

Interviews

- Do employees understand and accept the results of accident and near-miss investigations?
- Do employees mention a tendency on management's part to blame the injured employee?
- Do employees believe that all hazards contributing to accidents are corrected or controlled?

Site Conditions and Root Causes of Hazards

Are accidents sometimes caused at least partly by factors that might also have contributed to previous near-misses that were not investigated or accidents that were too superficially investigated?

Injury and Illness Pattern Analysis

Documentation

- In addition to the required OSHA log, are careful records kept of first aid injuries and/or illnesses that might not immediately appear to be work-related?
- Is there any periodic, written analysis of the patterns of near-misses, injuries, and/or illnesses over time, seeking previously unrecognized connections between them that indicate unrecognized hazards needing correction or control?
- Looking at the OSHA 200 log and, where applicable, first aid logs, are there patterns of illness or injury that should have been analyzed for previously undetected hazards?
- If there is an occupational nurse/doctor on the worksite, or if employees suffering from ordinary illness are encouraged to see a nearby health care provider, are the lists of those visits analyzed for clusters of illness that might be work-related?

Interviews

Do employees mention illnesses or injuries that seem work-related to them but that have not been analyzed for previously undetected hazards?

Site Conditions and Root Causes of Hazards (Not generally applicable.)

3. ASSESSING THE KEY COMPONENTS OF HAZARD PREVENTION AND CONTROL

Appropriate Use of Engineering Controls, Work Practices, Personal Protective Equipment, and Administrative Controls

Documentation

- If there are documented comprehensive surveys, are they accompanied by a plan for systematic prevention or control of hazards found?
- If there is a written plan, does it show that the best method of hazard protection was chosen?
- Are there written safe work procedures?
- If respirators are used, is there a written respirator program?

Interviews

- Do employees say they have been trained in and have ready access to reliable, safe work procedures?
- Do employees say they have difficulty accomplishing their work because of unwieldy controls meant to protect them?
- Do employees ever mention personal protective equipment, work procedures, or engineering controls as interfering with their ability to work safely?
- Do employees who use PPE understand why they use it and how to maintain it?
- Do employees who use PPE indicate that the rules for PPE use are consistently and fairly enforced?
- Do employees indicate that safe work procedures are fairly and consistently enforced?

Site Conditions and Root Causes of Hazards

- Are controls meant to protect workers actually putting them at risk or not providing enough protection?

- Are employees engaging in unsafe practices or creating unsafe conditions because rules and work practices are not fairly and consistently enforced?
- Are employees in areas designated for PPE wearing it properly, with no exceptions?
- Are hazards that could feasibly be controlled through improved design being inadequately controlled by other means?

Facility and Equipment Preventive Maintenance

Documentation

- Is there a preventive maintenance schedule that provides for timely maintenance of the facilities and equipment?
- Is there a written or computerized record of performed maintenance that shows the schedule has been followed?
- Do maintenance request records show a pattern of certain facilities or equipment needing repair or breaking down before maintenance was scheduled or actually performed?
- Do any accident/incident investigations list facility or equipment breakdown as a major cause?

Interviews

- Do employees mention difficulty with improperly functioning equipment or facilities in poor repair?
- Do maintenance employees believe that the preventive maintenance system is working well?
- Do employees believe that hazard controls needing maintenance are properly cared for?

Site Conditions and Root Causes of Hazards

- Is poor maintenance a frequent source of hazards?
- Are hazard controls in good working order?
- Does equipment appear to be in good working order?

Establishing a Medical Program

Documentation

Are good, clear records kept of medical testing and assistance?

Interviews

- Do employees say that test results were explained to them?
- Do employees feel that more first aid or CPR-trained personnel should be available?
- Are employees satisfied with the medical arrangements provided at the site or elsewhere?
- Does the occupational health care provider understand the potential hazards of the worksite, so that occupational illness symptoms can be recognized?

Site Conditions and Root Causes of Hazards

- Have further injuries or worsening of injuries occurred because proper medical assistance (including trained first aid and CPR providers) was not readily available?
- Have occupational illnesses possibly gone undetected because no one with occupational health specialty training reviewed employee symptoms as part of the medical program?

Emergency Planning and Preparation

Documentation

Are there clearly written procedures for every likely emergency, with clear evacuation routes, assembly points, and emergency telephone numbers?

Interviews

When asked about any kind of likely emergency, can employees tell you exactly what they are supposed to do and where they are supposed to go?

Site Conditions and Root Causes of Hazards

- Have hazards occurred during actual or simulated emergencies due to confusion about what to do?
- In larger worksites, are emergency evacuation routes clearly marked?
- Are emergency telephone numbers and fire alarms in prominent, easy to find locations?

4. ASSESSING THE KEY COMPONENTS OF SAFETY AND HEALTH TRAINING

Ensuring that all Employees Understand Hazards

Documentation

- Does the written training program include complete training for every employee in emergency procedures and in all potential hazards to which employees may be exposed?
- Do training records show that every employee received the planned training?
- Do the written evaluations of training indicate that the training was successful, and that the employees learned what was intended?

Interviews

- Can employees tell you what hazards they are exposed to, why those hazards are a threat, and how they can help protect themselves and others?
- If PPE is used, can employees explain why they use it and how to use and maintain it properly?
- Do employees feel that health and safety training is adequate?

Site Conditions and Root Causes of Hazards

- Have employees been hurt or made ill by hazards of which they were completely unaware, or whose dangers they did not understand, or from which they did not know how to protect themselves?

- Have employees or rescue workers ever been endangered by employees not knowing what to do or where to go in a given emergency situation?
- Are there hazards in the workplace that exist, at least in part, because one or more employees have not received adequate hazard control training?
- Are there any instances of employees not wearing required PPE properly because they have not received proper training? Or because they simply don't want to and the requirement is not enforced?

Ensuring that Supervisors Understand their Responsibilities

Documentation

Do training records indicate that all supervisors have been trained in their responsibilities to analyze work under their supervision for unrecognized hazards, to maintain physical protections, and to reinforce employee training through performance feedback and, where necessary, enforcement of safe work procedures and safety and health rules?

Interviews

- Are supervisors aware of their responsibilities?
- Do employees confirm that supervisors are carrying out these duties?

Site Conditions and Root Causes

Has a supervisor's lack of understanding of safety and health responsibilities played a part in creating hazardous activities or conditions?

Ensuring that Managers Understand their Safety and Health Responsibilities

Documentation

- Do training plans for managers include training in safety and health responsibilities?

- Do records indicate that all line managers have received this training?

Interviews

Do employees indicate that managers know and carry out their safety and health responsibilities?

Site Conditions and Root Causes of Hazards

Has an incomplete or inaccurate understanding by management of its safety and health responsibilities played a part in the creation of hazardous activities or conditions?

CONCLUSION

The key to a successful and efficient evaluation is to combine elements when using each technique. First review the documentation available relating to each element. Then walk through the worksite to observe how effectively what is on paper appears to be implemented. While walking around, interview employees to verify that what you read and what you saw reflects the state of the safety and health program.

Effective safety and health program evaluation is a dynamic process. If you see or hear about aspects of the program not covered in your document review, ask to receive the documents, if any, relating to these aspects. If the documents included program elements not visible during your walk around the site and/or not known to employees, probe further. Utilizing this cross-checking technique should result in an effective, comprehensive evaluation of the worksite's safety and health program.

Appendix C

Draft Safety and Health Program Rule

DRAFT PROPOSED SAFETY AND HEALTH PROGRAM RULE
29 CFR 1900.1
Docket No. S&H-0027

<u>What is the purpose of this rule?</u> The purpose of this rule is to reduce the number of job-related fatalities, illnesses, and injuries. The rule will accomplish this by requiring employers to establish a workplace safety and health program to ensure compliance with OSHA standards and the General Duty Clause of the Act (Section 5(a)(1)).

(a) Scope.

(a)(1) <u>Who is covered by this rule?</u> All employers covered by the Act, except employers engaged in construction and agriculture, are covered by this rule.

(a)(2) <u>To what hazards does this rule apply?</u> This rule applies to hazards covered by the General Duty Clause and by OSHA standards.

(b) Basic obligation.

(b)(1) <u>What are the employer's basic obligations under the rule?</u> Each employer must set up a safety and health program to manage workplace safety and health to reduce injuries, illnesses and fatalities by systematically achieving compliance with OSHA standards and the General Duty Clause. The program must be appropriate to conditions in the workplace, such as the hazards to which employees are exposed and the number of employees there.

(b)(2) <u>What core elements must the program have?</u> The program must have the following core elements:

(i) Management leadership and employee participation;
(ii) Hazard identification and assessment;
(iii) Hazard prevention and control;
(iv) Information and training; and
(v) Evaluation of program effectiveness.

(b)(3) Does the rule have a grandfather clause? Yes. Employers who have implemented a safety and health program before the effective date of this rule may continue to implement that program if:
(i) The program satisfies the basic obligation for each core element; and
(ii) The employer can demonstrate the effectiveness of any provision of the employer's program that differs from the other requirements included under the core elements of this rule.

(c) Management leadership and employee participation.

(c)(1) Management leadership.

(c)(1)(i) What is the employer's basic obligation? The employer must demonstrate management leadership of the safety and health program.

(c)(1)(ii) What must an employer do to demonstrate management leadership of the program? An employer must:
(A) Establish the program responsibilities of managers, supervisors, and employees for safety and health in the workplace and hold them accountable for carrying out those responsibilities;
(B) Provide managers, supervisors, and employees with the authority, access to relevant information, training, and resources they need to carry out their safety and health responsibilities; and
(C) Identify at least one manager, supervisor, or employee to receive and respond to reports about workplace safety and health conditions and, where appropriate, to initiate corrective action.

(c)(2) Employee participation.

(c)(2)(i) What is the employer's basic obligation? The employer must provide employees with opportunities for participation in establishing, implementing, and evaluating the program.

(c)(2)(ii) What must the employer do to ensure that employees have opportunities for participation? The employer must:
(A) Regularly communicate with employees about workplace safety and health matters;
(B) Provide employees with access to information relevant to the program;
(C) Provide ways for employees to become involved in hazard identification and assessment, prioritizing hazards, training, and program evaluation;
(D) Establish a way for employees to report job-related fatalities, injuries, illnesses, incidents, and hazards promptly and to make recommendations about appropriate ways to control those hazards; and
(E) Provide prompt responses to such reports and recommendations.

(c)(2)(iii) What must the employer do to safeguard employee participation in the program? The employer must not discourage employees from making reports and recommendations about fatalities, injuries, illnesses, incidents, or hazards in the workplace, or from otherwise participating in the workplace safety and health program.

Note: In carrying out this paragraph (c)(2), the employer must comply with the National Labor Relations Act.

(d) Hazard identification and assessment.

(d)(1) What is the employer's basic obligation? The employer must systematically identify and assess hazards to which employees are exposed and assess compliance with the General Duty Clause and OSHA standards.

(d)(2) What must the employer do to systematically identify and assess hazards and assess compliance? The employer must:

(i) Conduct inspections of the workplace;

(ii) Review safety and health information;

(iii) Evaluate new equipment, materials, and processes for hazards before they are introduced into the workplace; and

(iv) Assess the severity of identified hazards and rank those that cannot be corrected immediately according to their severity.

Note: Some OSHA standards impose additional, more specific requirements for hazard identification and assessment. This rule does not displace those requirements.

(d)(3) How often must the employer carry out the hazard identification and assessment process? The employer must carry it out:

(i) Initially;

(ii) As often thereafter as necessary to ensure compliance with the General Duty Clause and OSHA standards and at least every two years; and

(iii) When safety and health information or a change in workplace conditions indicates that a new or increased hazard may be present.

(d)(4) When must the employer investigate safety and health events in the workplace? The employer must investigate each work-related death, serious injury or illness, or incident (near-miss) having the potential to cause death or serious physical harm.

(d)(5) What records of safety and health program activities must the employer keep? The employer must keep records of the hazards identified and their assessment and the actions the employer has taken or plans to take to control those hazards.

Exemption: Employers with fewer than 10 employees are exempt from the recordkeeping requirements of this rule.

(e) Hazard prevention and control.

(e)(1) What is the employer's basic obligation? The employer's basic obligation is to systematically comply with the hazard prevention and control requirements of the General Duty Clause and OSHA standards.

(e)(2) If it is not possible for the employer to comply immediately, what must the employer do? The employer must develop a plan for coming into compliance as promptly as possible, which includes setting priorities and deadlines and tracking progress in controlling hazards.

> **Note:** Any hazard identified by the employer's hazard identification and assessment process that is covered by an OSHA standard or the General Duty Clause must be controlled as required by that standard or that clause, as appropriate.

(f) Information and training.

(f)(1) What is the employer's basic obligation? The employer must ensure that:

(i) Each employee is provided with information and training in the safety and health program; and

(ii) Each employee exposed to a hazard is provided with information and training in that hazard.

> **Note:** Some OSHA standards impose additional, more specific requirements for information and training. This rule does not displace those requirements.

(f)(2) What information and training must the employer provide to exposed employees? The employer must provide information and training in the following subjects:

(i) The nature of the hazards to which the employee is exposed and how to recognize them;

(ii) What is being done to control these hazards;

(iii) What protective measures the employee must follow to prevent or minimize exposure to these hazards; and

(iv) The provisions of applicable standards.

(f)(3) When must the employer provide the information and training required by this rule?

(f)(3)(i) The employer must provide initial information and training as follows:

(A) For current employees, before the compliance date specified in paragraph (i) for this paragraph (f); and

(B) For new employees, before initial assignment to a job involving exposure to a hazard.

> **Note:** The employer is not required to provide initial information and training in any subject in paragraph (f)(2) for which the employer can demonstrate that the employee has already been adequately trained.

(f)(3)(ii) The employer must provide periodic information and training:

(A) As often as necessary to ensure that employees are adequately informed and trained; and

(B) When safety and health information or a change in workplace conditions indicates that a new or increased hazard exists.

(f)(4) What training must the employer provide to employees who have program responsibilities? The employer must provide all employees who have program responsibilities with the information and training necessary for them to carry out their safety and health responsibilities.

(g) Evaluation of program effectiveness.

(g)(1) What is the employer's basic obligation? The employer's basic obligation is to evaluate the safety and health program to ensure that it is effective and appropriate to workplace conditions.

(g)(2) How often must the employer evaluate the effectiveness of the program? The employer must evaluate the effectiveness of the program:
(i) As often as necessary to ensure program effectiveness;
(ii) At least once within the 12 months following the final compliance date specified in paragraph (i); and
(iv) Thereafter at least once every two years.

(g)(3) When is the employer required to revise the program? The employer must revise the program in a timely manner to correct deficiencies identified by the program evaluation.

(h) Multi-employer workplaces.

(h)(1) What are the host employer's responsibilities? The host employer's responsibilities are to:
(i) Provide information about hazards, controls, safety and health rules, and emergency procedures to all employers at the workplace; and
(ii) Ensure that safety and health responsibilities are assigned as appropriate to other employers at the workplace.

(h)(2) What are the responsibilities of the contract employer? The responsibilities of a contract employer are to:
(i) Ensure that the host employer is aware of the hazards associated with the contract employer's work and what the contract employer is doing to address them; and
(ii) Advise the host employer of any previously unidentified hazards that the contract employer identifies at the workplace.

(i) Dates.

(i)(1) What is the effective date for this rule? The effective date for this rule is [insert date 90 days from the date of publication in the Federal Register].

(i)(2) When must the employer be in compliance with the requirements of this rule?
(i)(2)(i) Employers with fewer than 10 employees must comply with the requirements of paragraphs (c), (f), and (h) by [insert date 18 months after

the effective date], and with paragraphs (d), (e), and (g) by [insert date 36 months after the effective date].

(i)(2)(ii) Employers with 10 employees or more must comply with the requirements in paragraphs (c), (f), and (h) by [insert date 9 months after the effective date], and with paragraphs (d), (e), and (g) by [insert date 18 months after the effective date].

(j) Definitions.

Control means to reduce exposure to hazards in accordance with the General Duty Clause or OSHA standards, including providing appropriate supplemental and/or interim protection, as necessary, to exposed employees. Prevention and elimination are the best forms of control.

Contract employer is an employer who performs work for a host employer at the host employer's workplace. A contract employer does not include an employer who provides incidental services that do not influence the workplace safety and health program, whose employees are only incidentally exposed to hazards at the host employer's workplace (e.g., food and drink services, delivery services, or other supply services).

Employee means all persons who are considered employees under the OSH Act, including temporary, seasonal, and "leased" employees.

Employer means all persons who are considered employers under the OSH Act.

Exposure (exposed) means that an employee in the course of employment is reasonably likely to be subjected to a hazard.

General Duty Clause means the General Duty Clause of the OSH Act, Section 5(a)(1), which states that "[e]ach employer . . . shall furnish to each of his employees employment and a place of employment which are free from recognized hazards that are causing or are likely to cause death or serious physical harm to his employees."

Host employer means an employer who controls conditions at a multi-employer worksite.

Multi-employer worksite means a workplace where there is a host employer and at least one contract employer.

Program means procedures, methods, processes, and practices that are part of the management system at the workplace.

Safety and health information means the establishment's fatality, injury, and illness experience, OSHA 200logs, workers' compensation claims, nurses' logs, the results of any medical screening/surveillance, employee safety and health complaints and reports, environmental and biological exposure data, information from prior workplace safety and health inspections, Materials Safety Data Sheets (MSDSs), the results of employee symptom surveys, safety manuals and health and safety warnings provided to the employer by equipment manufacturers and chemical suppliers, information about occupational safety and health provided to the employer by trade associations or professional safety or health organizations, and the results of prior accident and incident investigations at the workplace.

Severity means the likelihood of employee exposure, the seriousness of harm associated with the exposure, and the number of exposed employees.

Appendix D

Example Inspection Form

PROJECT SAFETY INSPECTION REPORT

PROJECT _____ **DATE** _____

	YES	NO	N/A

FIRST AID

1. Are first aid kit locations identified and accessible? _____ _____ _____
2. Are emergency eye wash/safety showers available and inspected monthly? _____ _____ _____
3. Are first aid kits inspected weekly? _____ _____ _____
4. Is a qualified first aid/CPR provider on site? _____ _____ _____

PERSONAL PROTECTIVE EQUIPMENT

1. Have levels of personnel protection been established? _____ _____ _____
2. Are respirators decontaminated, inspected, and stored according to standard procedures? _____ _____ _____
3. Have employees been fit-tested? _____ _____ _____
4. Is defective personal protective equipment tagged and taken out of service? _____ _____ _____
5. Does compressed breathing air meet CGA Grade "D" minimum? _____ _____ _____
6. Are there sufficient sizes and quantities of protective equipment? _____ _____ _____
7. At a minimum, are employees utilizing safety glasses, hard hats, and steel toe boots? _____ _____ _____

FIRE PREVENTION

1. Are employees smoking only in designated outdoor areas?
2. Are fire lanes established and maintained?
3. Are flammable liquid dispensing systems bonded?
4. Are approved safety cans available for storage of flammable liquids?
5. Has the local fire department been contacted?
6. Are fire extinguishers available and inspected monthly?
7. Are flammables and combustibles properly stored?
8. Are flammable storage cabinets available and used when needed?
9. Have non-flammables such as cardboard been removed from flammable cabinets?
10. Are flammable storage cabinets grounded?

AIR MONITORING

1. Is required air monitoring being conducted?
2. Are air monitoring instruments calibrated daily?
3. Are air monitoring logs up to date?
4. Are instrument user manuals available?
5. Are instruments being maintained?
6. Are employees notified of personal sampling results within 5 days of receipt?

WELDING AND CUTTING

1. Are fire extinguishers present at welding and cutting operations?
2. Are confined spaces evaluated prior to and during cutting and welding operations?
3. Have Hot Work Permits been completed?
4. Are proper helmets, goggles, aprons, and gloves available for welding and cutting operations?
5. Are welding machines properly grounded?

6. Are oxygen and fuel gas cylinders stored a minimum of 20 feet apart? ____ ____ ____
7. Are only trained personnel permitted to operate welding and cutting equipment? ____ ____ ____
8. Are gas cylinders transported in a secured vertical position with caps in place? ____ ____ ____

HAND AND POWER TOOLS

1. Are defective hand and power tools tagged and taken out of service? ____ ____ ____
2. Is eye protection available and used when operating power tools? ____ ____ ____
3. Are guards and safety devices in place on power tools? ____ ____ ____
4. Are power tools inspected before each use? ____ ____ ____
5. Are nonsparking tools available when necessary? ____ ____ ____
6. Is the correct tool being used for the job? ____ ____ ____

MOTOR VEHICLES

1. Are vehicles regularly inspected? ____ ____ ____
2. Are personnel licensed for the vehicles they operate? ____ ____ ____
3. Are unsafe vehicles tagged and reported to supervision? ____ ____ ____
4. Is vehicle's safety equipment operating properly? ____ ____ ____
5. Are loads secure? ____ ____ ____
6. Are vehicle occupants using safety belts? ____ ____ ____
7. Are current insurance cards and blank accident report forms located in vehicles? ____ ____ ____

EMERGENCY PLANS

1. Are emergency telephone numbers posted? ____ ____ ____
2. Have emergency escape routes been designated? ____ ____ ____
3. Are employees familiar with the emergency signal? ____ ____ ____
4. Has the emergency route to the hospital been established and posted? ____ ____ ____
5. Is a vehicle on site that can transport injured employees to the hospital? ____ ____ ____

MATERIALS HANDLING

1. Are materials stacked and stored to prevent sliding or collapsing?
2. Are tripping hazards identified?
3. Are semi-trailers chocked?
4. Are fixed jacks used under semi-trailers?
5. Are riders prohibited on materials handling equipment?
6. Are approved manlifts provided for the lifting of personnel?
7. Are personnel in manlifts wearing approved fall protection devices?

FIRE/CHEMICAL PROTECTION

1. Has a fire alarm system been established?
2. Do employees know the location and use of all fire extinguishers?
3. Are fire extinguisher locations posted?
4. Are combustible materials segregated from open flames?
5. Have fire extinguishers been professionally inspected during the last year?
6. Are all pull stations and fire extinguishers unobstructed and easy to reach?
7. Is all storage at least 18 inches from sprinkler heads?
8. Are fire extinguishers visually inspected monthly?
9. Is chemical compatibility considered in the storage of hazardous chemicals?
10. Are acids and bases stored separately in a corrosive cabinet?

ELECTRICAL

1. Is electrical equipment and wiring properly guarded and maintained in good condition?
2. Are extension cords kept out of wet areas?
3. Is damaged electrical equipment tagged and taken out of service?
4. Have underground electrical lines been identified by proper authorities?

5. Has a lockout/tagout system been established? _____ _____ _____
6. Are GFCIs being used on all temporary electrical systems and as needed? _____ _____ _____
7. Are extension cords being inspected daily (i.e., group pin in place, no unapproved splices)? _____ _____ _____
8. Are warning signs exhibited on high voltage equipment (250 V or greater)? _____ _____ _____
9. Is adequate distance maintained from overhead electrical lines? _____ _____ _____
10. Are switches, circuit breakers, and switchboards installed in wet locations enclosed in weatherproof enclosures? _____ _____ _____

CRANES AND RIGGING

1. Are cranes inspected daily prior to use? _____ _____ _____
2. Are crane swing areas barricaded or demarked? _____ _____ _____
3. Is all rigging equipment tagged with an identification number and rated capacity? _____ _____ _____
4. Is rigging equipment inspection documented? _____ _____ _____
5. Are slings, chains, and rigging inspected before each use? _____ _____ _____
6. Are damaged slings, chains, and rigging tagged and taken out of service? _____ _____ _____
7. Are slings padded or protected from sharp corners? _____ _____ _____
8. Do employees keep clear of suspended loads? _____ _____ _____
9. Are rated load capacities and special hazard warnings posted on the crane? _____ _____ _____
10. Are the records of annual crane inspection available? _____ _____ _____
11. Have accessible areas within the swing radius of the rear of the crane been barricaded? _____ _____ _____
12. Do crane operators have required training/ certification? _____ _____ _____

COMPRESSED GAS CYLINDERS

1. Are breathing air cylinders charged only to prescribed pressures? _____ _____ _____
2. Are like cylinders segregated and stored in well ventilated areas? _____ _____ _____
3. Is smoking prohibited in cylinder storage areas? _____ _____ _____
4. Are cylinders stored secure and upright? _____ _____ _____
5. Are cylinders protected from snow, rain, etc.? _____ _____ _____
6. Are cylinder caps in place before cylinders are moved? _____ _____ _____
7. Are fuel gas and oxygen cylinders stored a minimum of 20 feet apart? _____ _____ _____
8. Are propane cylinders stored and used only outside of buildings? _____ _____ _____

SCAFFOLDING

1. Is scaffolding placed on a flat, firm surface? _____ _____ _____
2. Are scaffold planks free of mud, ice, grease, etc.? _____ _____ _____
3. Is scaffolding inspected before each use? _____ _____ _____
4. Are defective scaffold parts taken out of service? _____ _____ _____
5. Have employees completed scaffold user training? _____ _____ _____
6. On scaffolds where platforms are overlapped, is planking overlapped a minimum of 12 inches? _____ _____ _____
7. Does scaffold planking extend over end supports between 6 to 18 inches (dependent upon platform length)? _____ _____ _____
8. Are employees restricted from working on scaffolds during storms and high winds? _____ _____ _____
9. Are all pins in place and wheels locked? _____ _____ _____
10. Is required perimeter guarding (top rail, mid rail, and toe board) present? _____ _____ _____
11. Has a competent person been designated to oversee scaffold construction? _____ _____ _____
12. Are employees prohibited from moving mobile scaffold horizontally while employees are on them? _____ _____ _____

13. Are all scaffold components manufactured by the same company? _____ _____ _____

WALKING AND WORKING SURFACES

1. Are ladders regularly inspected? _____ _____ _____
2. Are accessways, stairways, ramps, and ladders clean of ice, mud, snow, or debris? _____ _____ _____
3. Are ladders being used in a safe manner? _____ _____ _____
4. Are ladders kept out of passageways, doors, or driveways? _____ _____ _____
5. Are broken or damaged ladders tagged and taken out of service? _____ _____ _____
6. Are metal ladders prohibited in electrical service? _____ _____ _____
7. Are stairways and floor openings guarded? _____ _____ _____
8. Are safety feet installed on straight and extension ladders? _____ _____ _____
9. Is general housekeeping being maintained? _____ _____ _____
10. Are ladders tied off? _____ _____ _____
11. Are handrails and siderails installed along the unprotected sides of stairways having 4 or more risers or rising more than 30 inches? _____ _____ _____

SITE SAFETY PLAN

1. Is a site safety plan available on site or accessible to all employees? _____ _____ _____
2. Does the safety plan accurately reflect site conditions and tasks? _____ _____ _____
3. Have potential hazards been described to employees on site? _____ _____ _____
4. Is there a designated safety official on site? _____ _____ _____
5. Have all employees signed the safety plan acknowledgment form? _____ _____ _____

SITE POSTERS

1. Are the following posters displayed in a prominent and accessible area?
   A. Minimum Wage _____ _____ _____
   B. OSHA Job Protection _____ _____ _____
   C. Equal Employment Opportunity _____ _____ _____

2. Are all required state-specific posters displayed? _____ _____ _____

SITE CONTROL

1. Are work zones clearly marked? _____ _____ _____
2. Are support trailers located to minimize exposure from a potential release? _____ _____ _____
3. Are support trailers accessible for approach by emergency vehicles? _____ _____ _____
4. Is the site properly secured during and after work hours? _____ _____ _____
5. Is an exclusion zone sign-in/sign-out log maintained? _____ _____ _____
6. Are only employees with current training and physicals permitted in exclusion zone? _____ _____ _____

HEAVY EQUIPMENT

1. Is heavy equipment inspected as prescribed by the manufacturer? _____ _____ _____
2. Is defective heavy equipment tagged and taken out of service? _____ _____ _____
3. Are project roads and structures inspected for load capacities and proper clearances? _____ _____ _____
4. Is heavy equipment shut down for fueling and maintenance? _____ _____ _____
5. Are backup alarms installed and working on mobile equipment? _____ _____ _____
6. Have qualified equipment operators been designated? _____ _____ _____
7. Are riders prohibited on heavy equipment? _____ _____ _____
8. Are guards and safety appliances in place and used? _____ _____ _____
9. Are operators using the "three point" system when mounting/dismounting equipment? _____ _____ _____

EXCAVATION

1. Has a "competent person" been designated to oversee excavation activities? _____ _____ _____
2. Prior to opening excavations, are utilities located and marked? _____ _____ _____
3. Has a professional engineer evaluated all excavations greater than 20 feet deep? _____ _____ _____

4. Is there rescue equipment on site and accessible to the excavation area? _____ _____ _____
5. Is excavated material placed a minimum of 24 inches from the excavation? _____ _____ _____
6. Are the sides of excavations sloped or shored to prevent cave-ins? _____ _____ _____
7. Have excavations greater than 4 feet deep been monitored for hazardous atmospheres (i.e., LEL/O_2)? _____ _____ _____
8. Are ladders or ramps used in excavations over 4 feet deep? _____ _____ _____
9. Are means of egress available so as to require no more than 25 feet of lateral travel? _____ _____ _____
10. Are barriers, i.e., guardrails or fences, placed around excavations near pedestrian or vehicle thoroughfares? _____ _____ _____
11. Is excavation inspected *daily* by competent persons and documented? _____ _____ _____

CONFINED SPACES

1. Have employees been trained in the hazards of confined spaces? _____ _____ _____
2. Are confined space permits posted at entrance to confined space? _____ _____ _____
3. Is a copy of the confined space entry procedure available? _____ _____ _____
4. Has a rescue plan been established? _____ _____ _____
5. Is an entry supervisor present at each permit-required entry? _____ _____ _____
6. Are required extraction/fall protection devices being used? _____ _____ _____

DECONTAMINATION

1. Are decontamination stations set up on site? _____ _____ _____
2. Is decontamination water properly contained and disposed of? _____ _____ _____
3. Are all pieces of equipment inspected for proper decontamination before leaving the site? _____ _____ _____
4. Are shin/metatarsal guards being used during power washing activities? _____ _____ _____

HAZARD COMMUNICATION

1. Is there a copy of the HAZCOM procedure on site?
2. Are there MSDSs for required materials/chemicals present on site?
3. Are all containers properly labeled, as to content, hazard?
4. Have employees been trained in accordance with the HAZCOM procedure?
5. Do employees (including subcontractors) know and understand the effects of exposure from the chemicals on site?
6. Have all personnel signed the HAZCOM acknowledgment form?
7. Is there an updated list of chemicals maintained on site?

TRAINING

1. Are tailgate safety meetings being conducted daily?
2. Are current training/medical records maintained on site?

DOCUMENTATION

1. Is an OSHA log maintained on site and posted in accordance with guidelines?
2. Are accident report forms available?
3. Is a copy of health and safety policy and procedures available on site?

PROJECT SAFETY INSPECTION REPORT

PROJECT: _____   **DATE:** _____

ALL NEGATIVE RESPONSES	CORRECTIVE ACTION	ASSIGNED TO	DATE ASSIGNED	DATE COMPLETED	VERIFIED BY

Appendix E

General Audit Form

Section One: General Communications

	YES	NO
Has the OSHA log been posted?		
Is the OSHA rights and responsibilities poster posted in a prominent place?		
Is the "No Lost Work Day" sign posted in a prominent place and kept up to date?		
Is the state-required Worker's Compensation poster prominently posted?		
Is there an in-house posting of whom to ask safety related questions or voice safety related concerns?		
Is there an in-house posting of where to report incidents and obtain incident report forms?		
Is the availability of a first-aid kit posted in the kitchen or well traveled area?		
Are copies of minutes from a local safety committee posted?		
Are upcoming safety related notices or events posted?		

Section Two: General Recordkeeping

	YES	NO
Are OSHA 200 and 300 logs available for review (past three years)?		
Are all personal accident-incident reports available for review?		
Do the incident reports match the OSHA logs?		
Are first-aid logs, close calls, and general incident reports available for review?		
Are equipment damage reports available for review?		
Are there investigation or follow-up reports available for all incidents?		

_____
_____
_____
_____
_____
_____

Section Three: Training Records

List all required training below. Then for each type of training determine if there is a record of completed training available for review (yes or no).

Training	YES	NO
Office orientation		
Field orientation		
Safe lifting		
Hazard communication		
Fire extinguisher		
Confined space		
Trenching/Excavation		
Supervisor training		
HAZWOPER initial training		
HAZWOPER refresher training		
HAZWOPER 3-day OJT		
HAZWOPER supervisor		
Safe driver training		
Asbestos (ACM) PM		
ACM Supervisor		
ACM Project Designer		
ACM Worker		
ACM Management Planner		
Lead Risk Assessor		
_____		
_____		
_____		
_____		
_____		
_____		
_____		

_____
_____
_____
_____
_____
_____
_____

Section Four: General Information

	YES	NO
Is there a copy of the Safety Manual or Safety Program available for review?		
Is there someone assigned to keep this Safety Manual up to date?		
Does this location have an active safety committee?		
Does this safety committee meet regularly and post meeting minutes?		
Is there someone at the facility that is involved with a local chapter of a safety organization?		
Is there a procedure and evidence that safety information is shared with local team members?		
Is there documentation demonstrating positive reinforcement for safety (i.e., awards or milestone celebrations)?		
Is there documentation for disciplinary action for failure to adhere to safety rules?		
Is there a library or listing of job safety analyses (JSAs) available?		
Does the number of job safety analyses mirror the types of tasks performed at this facility?		

_____
_____
_____
_____
_____
_____

Section Five: Inspections

	YES	NO
Are routine inspections ongoing?		
Is operating or production equipment inspected (and the findings? documented) prior to energization?		
Are results from inspections communicated and deficiencies corrected?		
Are vehicles inspected daily before use?		
Is documentation of vehicle pre-trip inspections available for review?		
Are ergonomics assessments being performed as per company requirements?		
Are recommendations from ergonomic assessments being completed promptly?		

_____
_____
_____
_____
_____
_____

Section Six: Medical Surveillance

	YES	NO
Has a medical provider been approved for this facility?		
Has management visited the medical facility and shared appropriate programs with the staff (such as the modified duty program, return to work slips, etc.)		
Are all team members medically cleared to perform assigned work tasks?		
Does all treatment/examination from the medical staff require a documented medical clearance form?		
Are all new hires required to take pre-employment medical exams prior to the start of work activity?		
Do all team members required to perform hazardous waste or other hazardous work activity have the appropriate timely medical clearance?		
Are all team members up-to-date for annual medical exams/clearance?		
Do all team members understand the procedure for obtaining injury treatment?		
Do all team members understand worker disposition procedures following an industrial accident?		
Do all team members understand the company's modified duty program?		
Do all team members required to wear respiratory complete respiratory training?		
Do all team members required to wear respiratory training have a timely record of a fit test?		

Appendix E: General Audit Form

	YES	NO
Are team members who use unique personal protective equipment medically cleared to do so?		
Are workers who require specific medical monitoring receiving this monitoring in a timely manner?		
Have the appropriate number of team members received First Aid/CPR training?		
Is the First Aid/CPR training up to date?		
Have the appropriate number of team members received bloodborne pathogen training?		
Is bloodborne pathogen protective equipment readily available?		

_____
_____
_____
_____
_____
_____
_____

Section Seven: Fire Prevention

	YES	NO
Does the facility have a fire extinguisher program in place?		
Does documentation indicate that fire extinguishers are being checked and maintained as required?		
Are team members trained (training documented) in the use of fire extinguishers?		
Are their pull stations placed near potential fire areas?		
Is there adequate clearance around extinguishers?		
Pull stations?		
Sprinkler heads?		
Is flammable material stored properly?		
Is there a flammable liquid storage cabinet being used to store flammable liquids?		
Is the flammable liquid storage cabinet being used in accordance with manufacturer recommendations and local fire codes:		
Grounding?		
Venting?		
Spark arrestors in or removed?		
Cardboard included in storage?		
Volume of stored material within limits?		
No spilled liquids present?		

	YES	NO
None other than flammables or non-compatible liquids being stored in cabinet?		
Fire houses, hose houses, hose reels, and other firefighting equipment is available and checked on a routine basis?		
Flammable materials are being stored away from hazards such as:		
Electrical panels?		
Space heaters?		
Open flames?		
Is the dispensing of flammables stored in an adequate location?		
Do drums of flammables stored horizontally on racks for dispensing include dead man valves and drip pans placed appropriately on dispensing drums to catch drippage?		
Do vertical drums for dispensing utilize a pumping system including tubing and piping that is compatible with the liquid being dispensed?		
Are flammable drums bonded and grounded appropriately when being stored or being dispensed?		
When flammable liquids are dispensed or stored in a container other than the original container, are only containers that are approved for industrial use being utilized?		
Is appropriate bonding and grounding equipment available and being utilized when dispensing flammable liquids?		

	YES	NO

Are no-smoking signs placed in accordance with fire codes and other applicable regulations near flammables?

Fire extinguishers:

Mounted?

Checked monthly and documented?

Have a location sign associated with them?

Has the use of the extinguisher been checked to ensure firefighting capability (i.e., A, B, C)?

Have team members been trained in the use of hot work permits?

Have projects been assessed for hot work type hazards such as open flames, cutting torches and welding, etc.?

Are past permits available for review?

_____
_____
_____
_____
_____
_____

Section Eight: Personal Protective Equipment (PPE)

	YES	NO
Does the PPE that is assigned to team members match the specification in the job safety analysis (JSA)?		
Is the PPE that the team member is utilizing adequate to match the hazards?		
Is enough PPE available in stock?		
Are the sizes of PPE available adequate to fit team members who may be required to wear PPE?		

_____
_____
_____
_____
_____
_____

Section Nine: *Material Handling and Storage*

	YES	NO
Does the material handling equipment being utilized have an adequate capacity and fit the work and space requirements?		
Is the height and stability of storage appropriate?		
Do pallet racks and shelves have the capacity to hold their load?		
Is the capacity posted on the storage rack?		
Is appropriate equipment available to reach top shelf storage safely (portable stairs, ladders, forklifts, etc.)?		
For forklifts:		
Trained team members?		
Adequate pre-trip record?		
Adequate maintenance?		
Adequate capacity?		
Not too large in size so as to fit the job?		
Charging station clean, with signage, and eye wash/shower for corrosives if necessary?		

_____
_____
_____
_____
_____
_____
_____

Section Ten: Tools and Equipment

Are first aid kits:

	YES	NO
Stocked?		
Checked?		
Documented?		
Are eye wash/safety showers inspected in a timely manner?		
Are the inspections documented?		
Are fire exits inspected regularly?		
Are the inspections documented?		
Are emergency lights inspected regularly?		
Are the inspections documented?		

_____
_____
_____
_____
_____

Index

accident costs 13
accident investigation(s) 6, 37, 233
accident prevention 36
accident prone 120
accident trend 33, 151
assessments (general) 28, 41

back problems (injury) 207, 208, 210, 212, 213
background checks 52
behavior based safety 15
blame game 226
bloodborne pathogen 95, 97

cash awards 133
common sense 180, 208
company physician 240
confined space 91
control group 16
culture 1
culture gap 25
cumulative trauma disorder (ctd) 138

Deming philosophy 15
direct cost 26
drug screening 63
   failure (drug test) 66
   for cause testing 67, 68
   pre employment testing 67
   random testing 67

employee accident reports 36
employee participation 159
Environmental Protection Agency (US EPA) 138

excavation trenching 95, 96
experience modification rate 154

field audits 171
first aid logs 35, 36
front line supervisor (fls)—field supervisor 37, 106, 127, 128, 129, 185, 186
follow up screening 59
Form 45's 36

gotcha 229, 248
group safety milestones 133, 135

Hawthorne effect 15, 62, 81, 106, 113, 132, 137, 186, 217, 220, 247
hazard prevention and control 87, 161
hazard communication program 147
hearing conservation 95, 97
hearing loss 211, 212
housekeeping 173

incentive awards, plans programs 61, 134, 136, 137
incident rate 149
incident logs 35
inspections 171, 197, 254

job safety (hazard) analysis 6, 25, 39, 85, 87, 108, 143, 144, 145, 146, 147, 150, 184, 245, 246

late reporting (incidents 235, 236
leadership 11, 219, 220
lockout tagout 95

Lower explosive limit (LEL) 142
lucrative bonus programs 127, 132, 251, 252

machoism 184, 244
management leadership 159
mandatory drug screening 68
mandatory training 93
Maslow's Hierarchy 71
material safety data sheets 144, 147
medical treatment 240
Monday morning cases 32
multilanguage employment applications 49

new hire 45, 64, 72
non-serious injury 34
Nuclear Regulatory Commission 61

obstacles 162
occupational physician 56
orientation 72, 88
OSHA (OSHA log) 6, 8, 29, 31, 32, 34, 35, 40, 42, 55, 86, 106, 108, 114, 138, 139, 142, 143, 148, 159, 161, 168, 170,
OSHA citations 39
OSHA recordability 30

performance appraisal—evaluations 111, 112, 122, 123, 125
permissible exposure limit 142
personal protective equipment (PPE) 4, 5, 59, 79, 94, 95, 96, 97, 129, 184, 208, 245, 246, 247
personal and confidential 207
personal physician 238
physician 38
physicals 61
post accident investigation 237
pre-employment (pre-placement) physical 56, 58
prescription drugs 35
pre-trip inspections 173, 174
productivity 16, 216

recordability 34, 35, 153
reference checks 52
respiratory protection 95, 96
review boards 200, 201
root cause 42

safe driving 95
safety audits 39, 41, 188, 18, 190, 191, 194, 195, 196, 197, 200, 202, 204, 254, 255
safety celebrations 257
safety culture 8, 37, 38, 39, 75, 151, 216
safety incentive (bonus) plan 126, 250, 251, 254
safety inspections 39, 41, 172
safety meetings 439, 1
Safety Observer Program 16
safety program 147, 172
safety task analysis 85
salary review 111
serious injury 34
safety performance 251
safety recognition 256
safety service awards 256
site specific safety plan 143, 147, 172
STAR program 31
steering committee 204, 209, 210
supervisor accident reports 36
supervisors (Front Line, Field) 215, 217, 218, 219, 220, 221, 222, 223, 224, 225, 226, 227 , 228, 229, 230 , 231, 233, 237, 238, 239, 240, 241, 242, 247, 248, 249
support 11

teaming 107
tend determination 154
test group 16
Total Quality Management 15, 20
training (safety and health 83, 85, 87, 88, 92, 94, 96, 161
training (regulation driven) 94
training (competent person) 96

Index

U.S. Department of Transportation (DOT) 56, 68, 69
underground storage tanks (UST) 94, 96, 97, 98
universal training 92
unsafe acts 10, 25, 250

vision 11

walking the talk 10
Workers' Compensation 8, 36, 37, 154, 211

working within medical restriction 241
workplace violence 62
worksite analysis 160

x rays 61
X, Y, Z theory 18, 19, 20, 35, 220

zero accidents (incidents) 8, 151